The
Ninth (Queen's Royal) Lancers

Engr. J. J. Whittington

Her Most Gracious Majesty
Queen Adelaide.

The
Ninth (Queen's Royal) Lancers

1715-1903

BY

FRANK H. REYNARD

LATE CAPTAIN NINTH LANCERS

DEDICATED

TO THE

OFFICERS, NON-COMMISSIONED OFFICERS, AND MEN;

WHO HAVE SERVED, AND ARE SERVING,

IN THE REGIMENT.

PREFACE

I WAS asked by Major-General H. A. Bushman, C.B. (Colonel of the Regiment), to try and compile an authentic account of the doings thereof from the date of its formation to the present time, and the following pages are the result.

I trust they may be of interest, not only to those who have served or are serving in the regiment, but also to their relatives and friends; and possibly to the general public.

The regimental official records have been taken as the basis of the work, amplified by much information from papers in the Public Record Offices of London and Dublin, from authors of acknowledged repute,[1] as also from many private sources.

[1] 'History of the War in the Peninsula,' Lieut.-General Sir Wm. Napier, K.C.B.; 'Narrative of the Second Sikh War, 1848-49,' E. T. Thackwell; 'The Sikhs and the Sikh Wars,' General Sir C. Gough and A. D. Innes; 'Decisive Battles of India,' Malleson; 'History of the Indian Mutiny,' Kaye and Malleson; 'The Sepoy Revolt,' General M'Leod Innes, V.C.; 'Life of General Sir Hope Grant,' Colonel Knollys; 'Forty-one Years in India,' Lord Roberts, V.C.; 'Standards and Colours of the Army, 1661-1881,' S. M. Milne; 'A History of the Dress of the British Soldier,' Lieut.-Colonel T. Luard; 'Historical Records of the British Army,' Richard Cannon.

I desire to offer my most sincere thanks to all who have kindly assisted me with the work, especially to Lieut.-Colonel F. Colvin and Captain E. R. Gordon, who have allowed me to make use of their diary, in compiling the account of the South African war.

CONTENTS

PAGE

VARIOUS TITLES OF THE REGIMENT xv

CHAPTER I

1715–1718

DETAILS CONCERNING THE FORMATION OF THE REGIMENT—THE JACOBITE REVOLT—BATTLE OF PRESTON—MOVE TO IRELAND 1

CHAPTER II

1719–1801

SERVICE IN IRELAND—RATES OF PAY—COST OF A CAVALRY REGIMENT—CONSTITUTED LIGHT DRAGOONS—THE IRISH REBELLION, WITH INCIDENTS 8

CHAPTER III

1803–1828

WAR WITH SPAIN—IN SOUTH AMERICA—CAPTURE OF MONTE VIDEO AND BUENOS AYRES—WALCHEREN EXPEDITION—PENINSULA CAMPAIGN—INTRODUCTION OF THE LANCE—DESIGNATED NINTH LANCERS—ISSUE OF REGIMENTAL STANDING ORDERS 19

CHAPTER IV

1829–1842

SERVICE AT HOME—APPOINTED (QUEEN'S ROYAL) LANCERS IN HONOUR OF QUEEN ADELAIDE 34

CHAPTER V

1842–1853

PAGE

Arrival in India—Actions of Punniar, Sabraon, Chillianwallah, and Goojerat—Proposal to give Lancers a Carbine, rejected 39

CHAPTER VI

1857–1859

A Brief Account of the most important Engagements in which the Regiment took part, during the Indian Mutiny 47

CHAPTER VII

1860–1885

Home Service to 1875—India, 1875—Afghan War, 1878-1880—Durbar, Rawal Pindi, 1885 . . . 78

CHAPTER VIII

1886–1899

Home Service—South Africa, 1896—India, 1898-1899 . 96

CHAPTER IX

1899–1903

India to South Africa—South African War, October 1899 to March 1902—Return to India—Delhi Coronaton Durbar, December 1902 . . 104

APPENDIX I

List of Battles (With Medals) 147

APPENDIX II

Guidons and Badges, Regimental Call, and Marches. 149

APPENDIX III

PAGE

Establishments 153

APPENDIX IV

Record of Service at Home and Abroad . . . 154

APPENDIX V

Alterations in Dress, Arms, &c. 158

APPENDIX VI

Chronological List of Colonels-Commanding (with Portraits and Memoirs) 164

Chronological List of Lieut.-Colonels Commanding . 180

Chronological List of Adjutants . . . 183

Chronological List of Quartermasters . . . 184

APPENDIX VII

Annual Army Lists. 1715–1903 185

APPENDIX VIII

Alphabetical List of Officers Appointed to the Regiment, 1715–1903. Showing Date of Appointment, Cause of Leaving, &c. . . . 231

APPENDIX IX

Annual List of Officers attending the Regimental Dinner, 1866–1903 246

INDEX 253

LIST OF PLATES

HER MAJESTY QUEEN ADELAIDE (*Photogravure*) *Frontispiece*

1. ONE OF WYNNE'S DRAGOONS, A.D. 1715 (*Coloured*) .
2. A 9TH LIGHT DRAGOON, A.D. 1804 (*Coloured*) . .
3. A 9TH LANCER, A.D. 1816 (*Coloured*) . . .
4. A 9TH LANCER, A.D. 1822 (*Coloured*) . . .
5. A 9TH LANCER, A.D. 1829 (*Coloured*) . . .
6. A 9TH (QUEEN'S ROYAL) LANCER, A.D. 1831 (*Coloured*) .
7. A 9TH (QUEEN'S ROYAL) LANCER, A.D. 1846-54 (*Coloured*)
8. A 9TH (QUEEN'S ROYAL) LANCER, A.D. 1896 (*Coloured*) .
9. MEDALS AND RIBBONS (*Coloured*) . . . .
10. GUIDONS OR STANDARDS (*Coloured*) . . .
11. MAJOR-GENERAL OWEN WYNNE (*Photogravure*) . .
12. GENERAL PHILIP HONYWOOD (*Photogravure*) . .
13. GENERAL SIR JAMES WALLACE SLEIGH (*Photogravure*) .
14. GENERAL SIR JAMES HOPE GRANT (*Photogravure*) .
15. GENERAL SIR ARCHIBALD LITTLE (*Photogravure*) .
16. GENERAL SIR WILLIAM DRYSDALE (*Photogravure*) .
17. MAJOR-GENERAL H. A. BUSHMAN (*Photogravure*) .
18. COLONEL CHARLES MORLAND (*Photogravure*) . .
19. LORD LOUGHBOROUGH (*Photogravure*) . . .

A.D. 1715.

A.D. 1804-1811.

A.D. 1816.

A.D. 1822.

A.D. 1829.

Officer—1831.

1846–1855.

A.D. 1896.

Medals—1811-1903.

Medals—1811-1903.

Medals—1811-1903.

Medals—1811-1903.

The King's, or First Guidon.
1743-1751.

The 2nd and 3rd Guidons.
1743-1751.

VARIOUS TITLES OF THE REGIMENT

July 1715 to 1719	. .	Wynne's Dragoons.
1719 to 1783	. .	9th Dragoons.

Note.—Although officially termed "9th Dragoons"; in all old Returns, Routes, Pay-Lists, Muster Rolls, &c., the regiment is always referred to by the name of its Colonel, at any rate up to 1780 or later.

1783 to Sept. 19, 1816	. .	9th Light Dragoons.
Sept. 19, 1816 to July 1830	.	9th Lancers.

Note.—By order of the Prince Regent of 19th Sept. 1816 the regiment was ordered to be equipped as Lancers, and from this date was so to be designated.

1830 to 1903	. .	9th (Queen's Royal) Lancers.

Note.—By notification in the *London Gazette*, 24th July 1830, H.M. King William IV. was pleased to direct that the regiment should assume the title of the 9th (Queen's Royal) Lancers, to be thus called after his Royal Consort, Queen Adelaide.

It seems probable that from this date the regiment assumed the cypher monogram of A.R. reversed, which it bears to the present day.

The

Ninth (Queen's Royal) Lancers

CHAPTER I

1715–1718

It may be well to briefly state the causes leading up to the formation of our regiment, now called the 9th (Queen's Royal) Lancers, but originally known as Wynne's Dragoons. Two regiments of Cuirassiers and fourteen of Dragoons, which had been raised for service during the reign of Queen Anne, were disbanded after the Peace of Utrecht, 1713; when the army was placed on a lower establishment. 1715.

During the years 1712 to 1714, there was apprehension lest the Stuart dynasty should be restored to the throne in the person of Queen Anne's half-brother; James Francis, Duke of Cornwall, the Chevalier de St. George, commonly known as "The Pretender."

Queen Anne died on the 1st August 1714 without issue, and under the Act of Settlement a great-grandson of James I., George, Elector of Hanover, succeeded to the throne; and was without opposition proclaimed sovereign of great Britain and Ireland, under the title of George I.

1715. The adherents of the Stuart dynasty had not been idle however, and were quietly working to place their representative, "The Pretender," on the throne. On the 20th July 1715 His Majesty (George I.) informed Parliament that a rebellion had begun at home, and that the nation was threatened with invasion from abroad. Measures to meet this danger were at once taken. A reward of £100,000 was offered to such person as should seize "The Pretender," dead or alive.

The want of more troops for defence of the crown and kingdom was felt at this crisis, and commissions were at once issued for the restoration of Major-General Pepper's Regiment (now 8th Hussars), and for the raising of seventeen additional regiments of Dragoons and thirteen of Foot.

The first of the new Dragoons thus embodied was our regiment, and the formation of it was intrusted to Major-General Owen Wynne, who had commanded the 5th Royal Irish Dragoons (5th Lancers) in 1695, and had, in the year 1705, raised a regiment of Foot for service under Queen Anne.

The following is a copy of the warrant whereby the raising of the regiment was authorized:—

GEORGE R.

WHEREAS we have thought fitt, that a Regiment of Dragoons be forthwith raised under your command for our Service, which is to consist of Six Troops; of one Sergeant, two Corporals, 1 Drummer, 1 Hautbois, and thirty Private Dragoons, including the Widow's men in each Troop. THESE are to authorize you by beat of drum or otherwise, to raise so many Voluntiers as shall be wanting to complete the said Regiment to the above numbers. And when you shall have listed 15 men fitt for service, in any of the said Troops, you are to give notice to two of our Justices of the Peace of the town or county where 1715. the same are, who are hereby authorized and required to view the said men, and certify the day of their so doing, from which day the said 15 men, and the Commission and Non-Commission Officers of six Troops are to enter into Our pay. And you are to cause the said Voluntiers to be raised and Levy'd as aforesaid; to march under the command of such Commission Officers as you shall direct, to Reading in Berkshire, appointed for the Rendezvous of the said Regiment; and all Magistrates, Justices of the Peace, Constables and other of our Officers whom it may concern, are hereby required to be assisting unto you, in providing quarters, impressing carriages, and otherwise, as there shall be occasion.

Given at Our Court at St. James's, 23rd July 1715, in the first year of Our reign.

By His Majesty's command.

WM. PULTENEY.

To our trusty and well beloved Owen Wynne, Major-General; or to the Officer or Officers appointed by him to raise Voluntiers for that Regiment.

It may here be of interest to note the careers of the seventeen Dragoon and thirteen Foot regiments at this date raised.

DRAGOONS

Wynne's	now 9th Lancers.	Bowles's,	now 12th Lancers.
Gore's	,, 10th Hussars.	Munden's,	,, 13th Hussars.
Honywood's,	,, 11th ,,	Dormer's,	,, 14th ,,

Total 6. Permanent.

Churchill's,	Stanhope's,
Sir Robert Rich's,	Molesworth's,
Tyrrell's,	Newton's,
Palmer's,	Lord Montjoy's,
Ferrer's,	Morris's,
Labouchetiere's,	(Total 11).

1715. Eleven of the Dragoon, and the thirteen regiments of Foot were all disbanded about 1717.

The regiment (Wynne's Dragoons) was raised in the southern counties of England, and recruiting seems to have been brisk, as its ranks were soon full. It was quartered at Bedford and other towns in the vicinity, and was soon in an efficient condition. It was at this time composed of six troops, the following being a list of the officers :—

Captains.	Lieutenants.	Cornets.
Colonel Owen Wynne.	Captain and Lieut. H. Crawford.	Owen Wynne.
Lieut.-Colonel H. Pearson.	William Witherington.	Wm. Carleton.
Major John Dunbar.	Edward Whitney.	Lewis Folliot.
Lord Leslie.	William Humphreys.	James Hill.
Andrew Knox.	Jacob Warnes.	Roger Pemberton.
Henry Smith.	Gustavus Hamilton.	Chris. Adams.

Note.—The senior lieutenant had the rank of captain, and commanded the colonel's troop.

Desertion from the army was not unknown at this period, as the following notice shows :—

NOTICE IN THE GAZETTE OF WHITEHALL

22nd December 1715.

(Edited by Samuel Buckley.)

Deserted out of Colonel Pearson's Troop, of Major-General Wynne's Regiment of Dragoons at Wigan, in Lancashire—Henry Dennison ; red-haired, round-ffaced, freckled, about 25 years of age, five ffoot eight inches high, born in Ireland, and by trade a taylor.

And also deserted from Captain Smith's Troop in the same Regiment, and from the same towne, Wm. Squire ; a middle-sized man, wears a brown wig, thin face, an Englishman, a carpenter by trade, and listed at Turnham Green in Middlesex.

Whosoever shall apprehend them, or either of them, and

give notice thereof to His Majesty's Secretary at Warr, at his office, Whitehall, so as that they may be brought to condign punishment, shall receive as a reward, the sum of 20s. for each of them. 1715.

WM. PULTENEY.

The friends of the Stuart dynasty had meantime not given up all hope; aid was supplied from France, and an armament was ready at Havre for "The Pretender's" service. The death of Louis XIV. on 1st September 1715 deprived them of foreign assistance. They had, however, proceeded too far in the scheme to retire with safety. The Earl of Mar raised the standard for "The Pretender" at Braemar on the 6th September, and was soon at the head of 10,000 men. He was opposed by the royal troops in Scotland, under the Duke of Argyll. At this period the regiment marched from Bedford to Derby and Manchester, where they arrived towards the end of October.

A portion of the rebel army had penetrated as far south as Lancashire, and was in possession of the town of Preston. General Wills, from Manchester, was ordered to at once take a force to oppose its further advance south. The regiments forming his force were—the 3rd or Pitt's Horse (2nd Dragoon Guards), Wynne's Dragoons (9th Lancers), Honywood's (11th Hussars), Munden's (13th Hussars), Dormer's (14th Hussars), Stanhope's (afterwards disbanded), and the 26th Regiment of Foot.

On the 12th November this force advanced on Preston. The town was defended by four barriers, but General Wills, being short of troops, decided to attack

1715. only two: those leading to Wigan and Lancaster, defended respectively by Colonel Mackintosh and Lord George Murray. He divided his force into three bodies—two for attack and one for support—as under:

1st Brigade. Brigadier Honywood (11th Hussars), in command of the Foot Regiments.

2nd Brigade. Brigadier Dormer, in command of Dormer's, Wynne's, and a squadron of Stanhope's Dragoons.

3rd Brigade. Brigadier Munden, in command of Pitt's Horse (2nd Dragoon Guards), Munden's, and a squadron of Stanhope's Dragoons.

1st Brigade. The 26th Regiment was directed to storm the approach from Wigan, and a detachment of fifty of Wynne's commanded by Captain Smith, a Lieutenant and a Cornet; and a like detachment from other Dragoon Regiments, the whole commanded by Lieutenant-Colonel Pearson of Wynne's, was ordered in support.

2nd Brigade. At the same time the approach from Lancaster was attacked by Wynne's and Dormer's Dragoons dismounted.

These forces advanced with great gallantry, but from want of cannon, were unable to finally force the entrenchments and barricades. The houses were set on fire, and a sharp musketry-fire was kept up.

On the 13th the attack was resumed, but on the 14th, reinforcements having arrived under Major-General Carpenter, the insurgents surrendered.

The loss of the regiment on this occasion was 6 men and 15 horses killed; 1 captain, 1 lieutenant,

1 cornet, 21 soldiers, and 15 horses wounded. The regiment after this was quartered at Warrington. 1715.

The regiment marched in April from Warrington to Winchester, and in September was reviewed near Petersfield (about twenty miles east of Winchester) by H.R.H. the Prince of Wales (afterwards George II.). 1716.

The regiment was employed on coast duty in Sussex, the establishment being 6 troops, 297 officers and men, and the upkeep was estimated at £12,849, 13s. a year. 1717.

Owing to the final suppression of the Stuart Rebellion, the strength of the army was this year reduced, and many regiments were disbanded: Wynne's Dragoons, however, was not one of them, and after receiving the thanks of the King for their general gallantry and good conduct, was ordered to proceed to Ireland. In June it marched to Bristol, and there embarked, coming on to the Irish establishment 25th June 1717.

In a marching order dated 31st May, it is recorded that Captain Douglas's troop "halt at Ballinrobe (if the inhabitants consent to quarter them there)." 1718.

CHAPTER II

1719–1801

1719. In July Major-General Owen Wynne was removed to the 5th Horse (4th Dragoon Guards), and was succeeded, 6th July, by Brigadier-General James Crofts (a natural son of the celebrated James, Duke of Monmouth). From this date the regiment was called the 9th Dragoons.

1720. Cornet Roger Pemberton, a corporal, and 14 men, by licence dated the 5th April 1720, proceeded to Great Britain to buy horses for the regiment.

1725. By licence dated the 13th April, 1 sergeant, 2 corporals, and 5 men, were ordered to Great Britain to bring horses over to Ireland.

1728. In June the regiment was reviewed at Athlone by the Right Honourable the Viscount Shannon.

1732. Major-General James Crofts having died in March, on the 29th May, Richard, Viscount Molesworth, was appointed colonel of the regiment. It was inspected at Sligo.

The regiment was inspected at Armagh. 1736.

On the 27th June, the Viscount Molesworth, having 1737. been appointed to the 5th Royal Irish Dragoons (5th Lancers), the colonelcy of the regiment was bestowed upon Brigadier-General John Cope, who had been colonel of the 5th Foot. In August 1741 he was appointed to the 7th Dragoons, and for a period of about nine months the colonelcy appears to have remained vacant.

The regiment was inspected at Longford. 1739.

On the 10th August, Lieutenant-Colonel John 1742. Brown, from the King's Regiment of Horse (1st Dragoon Guards), was appointed colonel. He only held the appointment for eleven months, and went to the 5th Horse (4th Dragoon Guards).

On 1st April, Colonel Henry de Grangue, from the 1743. 30th Foot, was appointed colonel, and having commanded for six years, accepted the colonelcy of the 4th Irish Horse (7th Dragoon Guards).

A copy of a muster-roll certificate of this date may be of interest.

ATHLONE, *5th June* 1743.

Mustered there in Captain Owen Wynne's Troop in His Majesty's Regiment of Dragoons, commanded by Major-General de Grangue, for three months, commencing the 1st day of April last past, and ending the last of June following, both included, being 91 days.

The Captain, Lieutenant, Cornet, and Quartermaster, with their led horses; one Sergeant, two Corporals, one Drummer,

1743. one Hautboy, with 21 Private Dragoons, 25 Troop horses; whereof the Hautboy absent on honour.

(Sd.) OWEN WYNNE.
JOHN CULLEN.
W. EMPSON. ST. LEGER ALDWORTH.

1746. On 22nd August, a licence was granted for Lieut. Edward Clayton and Quartermaster Adam Quartier, to proceed to Great Britain to buy horses for the regiment.

1749. On the 1st November, Lieut.-General George Reade, from colonel of the 9th Foot, was promoted to the colonelcy of the regiment, which he retained till his death in 1756.

1751. In 1747 a "Clothing Warrant" had been prepared by Colonel Robert Napier. In 1749 it was delivered by him, as "Adjutant-General," to the Clothing Board; and on the 1st July 1751 was officially promulgated, whereby changes were made in the dress of the regiment.

1756. On the death of Lieut.-General George Reade, Colonel John Jorden, from colonel of the 15th Foot, was appointed colonel of the regiment, dated 2nd April. This officer died in May.

On the 22nd May, Colonel Philip Honywood was appointed colonel from the 20th Foot, and in 1759 he was transferred to the 4th Horse (7th Dragoon Guards). This officer had a most gallant record of military service, and was a most able and valued public servant.

The following was the daily rate of pay and allow- 1757.
ances at this period :—

IRISH ESTABLISHMENT

Daily Rates of Pay

Rank.	Pay.	Subsistence.
Colonel	£1 11 4	£1 3 8
Lieutenant-Colonel	0 19 1	0 19 4
Major	0 17 11	0 17 4
Captain	0 12 4	0 9 2
Lieutenant	0 7 2	0 4 9
Cornet	0 6 2	0 3 10
Quartermaster	0 4 0	0 2 9
Chaplain	0 6 8	0 5 0
Surgeon	0 4 0	0 3 0
Surgeon's Mate	......	......
Drummer of Dragoons	0 1 8	0 1 0
Sergeant	0 2 8	0 1 7
Corporal	0 1 10	0 1 1
Hautboy	0 1 6	0 1 0
Private Man	0 1 6	0 0 11

Allowances per day in addition to the above :—

The colonel, 15s. a-day for clothing, etc., lost by deserters.

The colonel, the lieutenant-colonel, the major, and captains, a recruiting allowance either of 2s. 4d. a day, or that sum for each recruit. (Warrant, 14th August 1747.)

On 6th April, Colonel Henry Whitely, from the 1759.
10th Dragoons (10th Hussars), was promoted to the colonelcy of the regiment. He died as a lieutenant-general in 1770.

On 17th January, King George III. conferred the 1771.

1771. colonelcy of the regiment on Major-General James Johnston. In 1774 he was appointed Governor of Quebec, and in 1775 was transferred to the 1st Irish Horse (4th Dragoon Guards) in succession to his cousin, General James Johnston. This officer had a most distinguished career, military and civil, and was interred with great state at Westminster Abbey, 21st December 1779.

1775. On 27th April, Lieut.-Colonel Flower Mocher, from the 3rd Dragoon Guards, was appointed colonel, which position he held until his death in 1801, having been by that date promoted to the rank of general.

1778. The rate of pay of officers and men in this year was as follows :—

FIELD AND STAFF OFFICERS

	s.	*d.*		Per Year.		
Colonel, as Colonel, 12s. a-day, and 6 servants at 1s. 2d.	19	0	a-day	£346	15	0
Lieutenant-Colonel, as Lieutenant-Colonel	7	0	,,	127	15	0
Major, as Major	5	0	,,	91	5	0
Chaplain	6	8	,,	121	13	4
Adjutant	4	0	,,	73	0	0
Surgeon	4	0	,,	73	0	0
				£833	8	4

ONE TROOP

	s.	*d.*		Per Year.		
1 Captain, 10s. a-day, and 2 servants at 1s. 2d. each	12	4	a-day	£225	1	8
1 Lieutenant, 6s. a-day, and 1 servant at 1s. 2d.	7	2	,,	130	15	10
1 Cornet, 5s. a-day, and 1 servant at 1s. 2d.	6	2	,,	112	10	10

	s.	d.		Per Year.			
1 Quartermaster . . .	4	0	a-day	73	0	0	1778.
1 Sergeant . . .	2	8	„	48	13	4	
2 Corporals, 1s. 10d. each .	3	8	„	66	18	4	
1 Drummer . . .	1	8	„	30	8	4	
1 Hautboy . . .	1	6	„	27	7	6	
20 Private men, 1s. 6d. . .	30	0	„	547	10	0	
1 Corporal . . .	1	4½	„	25	1	10½	
10 Dismounted men, 1s. 0½d. .	10	5	„	190	2	1	
				£1477	9	9½	

Making the pay of a regiment of six troops £9698, 7s. 1d. a-year. Thus—

6 Troops at £1477, 9s. 9½d. each a-year .	£8864	18	9
Staff, for a year	833	8	4
	£9698	7	1

These rates of pay seem identical with those given for 1757, when it is remembered that the colonel, the lieutenant-colonel, and the major each commanded a troop and drew pay for such command, in addition to that of their ranks.

The following letter from the colonel of the regiment, who had evidently been called upon by the Paymaster-General to refund pay overdrawn, may be of interest:— 1780.

To His Excellency, John Earl of Buckingham, Lord-Lieutenant-General, and General Governor of Ireland.

The Memorial of Major-General Flower Mocher, Colonel of His Majesty's 9th Regiment of Dragoons, Humbly sheweth that several of the Officers, Non-Commission Officers and Private men of the said Regiment, who were absent at the

1780. Muster for six months ending 31st December 1778, stand chequed in the Muster Rolls of the said Regiment. That by deaths, desertions, and other accidents the same was unavoidable.

Your memorialist therefore prays Your Excellency's order to the Muster, Master General, for removing said Cheques.

Which is humbly submitted,

FLOWER MOCHER.

I approve.—J. IRVINE.

1783. The great value of light cavalry having now become apparent, the regiment was constituted a corps of Light Dragoons.

1784. On 18th May, by an Order from the Adjutant-General's office, Dublin, the clothing of Light Dragoons was altered from scarlet to dark blue.

1784-96. From 1768 to this date, the Government had found it necessary to maintain a large and efficient military force in Ireland, to overawe the rebellious spirit of the opponents of the Protestant succession. Although any hope of restoring the Stuart dynasty was dead, yet there was a secret antipathy to the Government; and on the breaking out of the French Revolution, the spirit of resistance in the country assumed a more decided aspect.

In December 1796 a French fleet arrived in Bantry Bay, but the Irish were not at the time prepared to rise.

1797. The military force in Ireland was however increased, and the regiment was raised to a strength of

8 troops; 31 officers, and 669 non-commissioned officers and men. In the summer it was encamped on the Curragh, and in September was inspected by Lieut.-General Sir David Dundas. 1797.

In this year the Irish malcontents received fresh promises of aid from France; and in May the Rebellion broke out. During the unnatural warfare which followed, the regiment was in the cause of duty actively employed, and frequently engaged against the insurgents. 1798.

On the 23rd May, a troop of the regiment quartered at Kilcullen (about ten miles from Kildare), under the command of Captain John Beevor, was suddenly attacked during the night. A party of the insurgents surrounded Captain Beevor's quarters; he was rescued by Corporals Swift and MacCuff. The detachment was speedily under arms at the "alarm-post"; Corporal Thomas King, a trumpeter, and 8 dragoons were surrounded by the insurgents before they could reach the alarm-post.

This party made a gallant dash; Corporal King and one dragoon cut their way through and joined the troop, the trumpeter and 7 dragoons were killed in the attempt. Quartermaster Charles King was taken prisoner by the insurgents, and was shot in their camp for persisting in his loyalty to his Sovereign.

Note.—It is uncertain whether Kilcullen was actually the scene of this incident.

On the 24th May, a troop of the regiment commanded by Captain the Honourable William Erskine, was engaged with a large body of the insurgents at

1798. Ballymore Eustace. The contest was severe: Captain Erskine, 3 sergeants, and 6 rank and file were killed; seven rank and file were wounded. The dragoons were finally successful, and the insurgents suffered severely.

On the same day Cornet Richard Love and 20 men of the regiment, together with 30 men of the Antrim Militia, advanced from Baltinglass to Stratford (near Carlow), where they met between 300 and 400 of the insurgents, whom they routed with a loss to themselves of a few wounded.

The headquarters of the regiment were at this time at Carlow, Lieut.-Colonel the Honourable Thomas Mahon commanding. There were also in the town a detachment of the 4th Dragoon Guards, a few Yeomanry, Militia, and Volunteers. Three thousand insurgents attempted to take the place, but were repulsed with a loss of about 400 men killed and many prisoners, among whom was their leader, Sir William Crosbie, Bart. He was tried by martial law and hanged. The casualties of the regiment were—2 dragoons killed, 2 sergeants, 1 trumpeter, and 9 dragoons wounded. The regiment received the thanks of the Governor in Public Orders, for their conduct on this occasion.

On the 5th June a troop of the regiment commanded by Captain John West Wilkins, together with a troop of the 5th Royal Irish Dragoons, formed a squadron under Captain William Irvine of the latter regiment at a town called New Ross, on the borders of County Kilkenny; Major-General Johnson being in command of that district. The insurgents attacked the town with great dash; and after a severe contest, during which the town was taken and retaken three times, they were finally utterly routed. Some sick and convalescent men under Assistant-Surgeon James Donaghoe of the regiment, greatly distinguished themselves, repulsing many attacks on the barracks, of which they had charge. The regiment had 4 rank and file killed; Captain Wilkins, 1 sergeant, 1 trumpeter, and 8 rank and file were wounded. 1798.

On the 18th June, four troops of the regiment were engaged, under Major-General Sir Charles Argill, Bart., in driving a body of the insurgents from the town of Castle Comer, and on the 19th a sharp action was fought at Kilcomney Hill (Kilcominey?). On these two occasions the regiment lost many men and horses.

The insurgents having established themselves at Vinegar Hill near Enniscorthy, 15,000 strong, dealt very hardly with hundreds of Protestants who fell into their hands. On the 21st June, two troops of the regiment under Captain Martin, formed part of a large force which attacked this place. The insurgents were routed and driven out: fourteen pieces of artillery, many small arms, and a quantity of plunder were taken: this engagement practically ending the Irish Rebellion, although on the 23rd June, a party under Lieut. James Lewes Higgins, together with some Yeomanry, was engaged at Leighlin with a body of insurgents who had escaped from Wexford, and on the 24th, a squadron of the regiment under Major Richard Donaldson, had a sharp encounter near Sharkill, on both of which occasions the insurgents suffered severely.

1798. The loss of the regiment was 3 rank and file killed; 2 sergeants, 1 trumpeter, 8 rank and file wounded.

The last survivor of the officers who served with the 9th in the Irish Rebellion, was Colonel Christopher Patrickson, who exchanged to the 43rd Light Infantry, and commanded it at Toulouse: he died September 1856.

A pay list of this date is here given:—

Pay List of the Regiment, April to September 1798.

Troops.	Total pay for Officers.			Total pay for N.C. Officers and Privates.			Total for Horses.		
The Colonel's . . .	£1511	13	9	£949	15	3	£201	9	8
,, Lieutenant Colonel's	416	6	6	964	17	7	203	15	0½
,, Major's . . .	382	0	3	947	5	3	176	2	3½
Captain Donaldson's .	339	6	3	939	10	3	196	10	2½
,, Beevor's . .	339	6	3	876	2	9	178	6	9
,, Erskine's . .	245	19	7	915	8	0	184	8	8
,, Martin's . .	339	6	3	942	9	0	192	19	7½
,, Needham's . .	339	6	3	883	4	0	218	13	10½
	£3913	5	1	£7418	12	1	£1552	6	1½

Total cost for six months, £12,984, 3s. 3½d.

This would probably include clothing, equipment, rations, and all allowances.

1801. On 1st August the King, George III., appointed Sir James St Clair Erskine, Bart., to be colonel of the regiment vice General Flower Mocher, who died this year. Sir James, among other appointments, had held that of Colonel of the Sussex Fencible Cavalry. In January 1805 he succeeded to the title of the Earl of Rosslyn.

CHAPTER III.

1803–1828

After a service of eighty-six years in Ireland, the regiment embarked at Dublin for England, and landed at Liverpool in May. It proceeded to Rumford, and from there to Ipswich, where it was brigaded with the 7th Light Dragoons (7th Hussars) under Major-General Lord Paget. 1803.

After so lengthened a period of service in Ireland, where it had never been together, and had not had the advantage of systematic supervision, its appearance, movements, and system were hardly up to date. By the zeal, activity, and *esprit de corps* of the regiment generally, this was soon remedied. On arrival in England a change was made in the equipment, as although constituted a Light Dragoon regiment in 1783, owing to its having been quartered in Ireland, it had retained some of its Heavy Dragoon appointments. From this year the colonel, the lieutenant-colonel, and major ceased to each have command of a troop.

During the summer of this year the regiment moved to Hounslow, and was employed on the King's duty (travelling escorts and letter-parties). 1804.

1805. In February it was augmented to ten troops—total, all ranks, 1100.

On the 10th June the regiment, together with the 14th Light Dragoons, was reviewed by George III. on Hounslow Heath. On the 14th it was, with the two regiments of Life Guards, two troops Horse Artillery, two brigades of Artillery, the 14th Light Dragoons, and two brigades of Foot Guards, again reviewed by the King on Wimbledon Common. About the 20th June, it moved from Hounslow to Guildford until September, when it proceeded to Blatchington and other parts of Sussex for coast duty.

1806. Early in this year the regiment moved to Hythe and Romney, and later to Ramsgate, Sandwich, and other places in Kent.

Following on diverse political events, war between Spain and Great Britain broke out. The Spanish possessions in South America were attacked, and in June, Buenos Ayres was captured by a small force under Major-General Beresford. The place was re-taken by the Spanish, and the English force made prisoners of.

A second attack upon the city was decided upon, and the regiment was selected to form part of the force detailed for this purpose. Eight troops, having transferred their horses to various regiments, and sending two troops to Coventry (presumably as a depôt), proceeded to Portsmouth in October, where they embarked (dismounted) for service in South America under Brigadier-General Sir Samuel Auchmuty. The expedition proceeded as far as Falmouth, and after lying in Falmouth Harbour some weeks, it got under weigh and sailed into the Channel. Having arrived so far (!), it was there discovered that the vessels in which the regiment was afloat were not fit for a long voyage. These put back to Falmouth, and the regiment being placed on board more serviceable ships, sailed in November for the Cape of Good Hope. It is recorded that the fleet touched at Port Praya in the island of St. Jago, to obtain water, and from that place the regiment continued the voyage alone, under escort of the frigate *Nereide*, reaching the river La Plata either the end of 1806 or January 1807. 1806.

Brigadier-General Sir Samuel Auchmuty landed in January at Maldonado, a town on the left bank of the river La Plata, having with him the 17th Light Dragoons, a detachment 21st Light Dragoons, some Royal Artillery and Royal Engineers, and the 40th, 87th, and 95th Regiments. This place together with the island of Goretti, had been taken possession of by the 38th and 47th Regiments under Colonel Backhouse, and the Naval Squadron under Sir Home Popham, in October 1806. On the 3rd February, Monte Video a town on the north side of the river, was taken by our troops. 1807.

The regiment disembarked on the 7th March, two troops proceeding to Colonia, the other six eventually to the town of Monte Video. Much sickness prevailed, and the Spaniards endeavoured to induce our men to desert.

At this time the cavalry at Monte Video were *the regiment*, the 17th Light Dragoons, and four troops of

1807. the 20th Light Dragoons (all dismounted). No exertions were made to mount the men, although the country was most suitable for the employment of cavalry. Eventually two squadrons of the 17th Light Dragoons and a troop of the 20th Light Dragoons received horses.

In May, large reinforcements arrived at Monte Video from the Cape of Good Hope; Lieut.-General Whitelocke who arrived at the same time, assuming command of the troops.

On the 28th June, an attack was made on the town of Buenos Ayres, a landing having been previously effected at Ensenada de Barragon. The regiment served on this occasion dismounted, and, with four troops of the 6th Dragoon Guards (Carabiniers), also dismounted, and the 40th and 45th Regiments of Foot, formed a brigade under the command of Colonel the Honble. Thos. Mahon (9th Light Dragoons). After some fatiguing marches the brigade arrived in the suburbs of Buenos Ayres, and the regiment was posted near the central avenue of the town. An attack was made on the 5th July, the four troops of the Carabiniers and three troops of the regiment attacking with two six-pounders, one of the main streets. The other five troops of the regiment were held in reserve in the centre.

The attack on this town does not appear to have been skilfully planned or well carried out, although the behaviour of the men, and their gallantry and dash, seem to have been quite up to the traditions of the army. Our loss was—2500 men killed, wounded, and taken prisoners.

The regiment's causalties were—Vety.-Surgeon Landers and 3 dragoons killed; Lieut. Cornwall, 2 sergeants, 1 trumpeter, and 21 rank and file wounded; also several rank and file missing. 1807.

Lieut.-General Whitelocke abandoned the object of the expedition, acceded to a treaty, and vacated the place. He was afterwards tried by a general court-martial and cashiered.

The regiment moved to Monte Video and embarked for England, arriving at Portsmouth in November. On the voyage home, one of the transport ships (the *James and Rebecca*), having got separated from the convoy, was wrecked in Mounts Bay, on the coast of Cornwall. This vessel had two troops on board—those of Captains Rathbone and Savage, with whom were Lieuts. Handley and Horsburg. The vessel struck at 11 P.M., and went to pieces at 11 A.M. the following day. Discipline, order, and method were well maintained. By aid of a hawser and traveller, which was established between the ship and the shore through the help of the Cornish people, the troops all landed safely, with the exception of Sergeant Hall and twenty-eight dragoons, who were drowned. The arms, baggage, accoutrements, and clothing of the troops were lost, and most of the men arrived ashore stark naked.

The regiment on return from South America, went to Guildford; recruiting parties being sent to Bath and Bristol, at which places many recruits were enlisted.

In the month of June the regiment moved to Hounslow, and entered upon the King's duty (escorts, orderlies, letter-carriers). 1808.

1809. In the summer of this year, the Government equipped a force for the attack of Holland, and six troops of the regiment (leaving four at home) were selected to take part in the expedition. They embarked at Portsmouth in July, the army being under the command of General the Earl of Chatham.

A landing was effected on the island of Walcheren near the mouth of the river Scheldt, and the town of Flushing was besieged. The regiment being the only cavalry there, was actively engaged during the siege, and after the capture of the town was stationed on the island of Walcheren. The expedition was finally abandoned, and the force returned to England. Four troops landed at Plymouth in October and proceeded to Exeter; two remained on the island till December, when they returned to England and joined the regiment. A most pestilential fever prevailed among the men while on this expedition, and the regiment lost between 150 and 200 men from "Walcheren fever."

1811. In the summer of this year six troops of the regiment were selected to join a British force under Lord Wellington, who was helping the Spanish and Portuguese against Bonaparte. Four of them embarked at Plymouth, and arrived at Lisbon in July; the other two followed, leaving four in England.

A short halt was made at Lisbon to refresh men and horses after their sea voyage. The regiment then marched to the Alentejo and joined a force there, which consisted of a division of cavalry, 10,000 infantry, and four brigades of artillery, under Lieut.-General

Rowland Hill; being brigaded with the 13th Light 1811.
Dragoons and the 2nd Hussars, King's German Legion (Brigadier, Major-General Long).

In October the force under Lieut.-General Hill was ordered to advance from Portalegre (about 100 miles E.N.E. of Lisbon) towards the Spanish frontier. After long marches in cold, wet, and stormy weather; at daybreak on the 28th it surprised a body of French troops under General Girard, which was halted at a village called Aroyo de Molino. Our approach was made under cover of a violent storm of rain and a thick mist. The French infantry was assembling outside the village to begin their march, the baggage was being loaded, and General Girard was waiting at his quarters for his horse. Suddenly the 71st and 92nd Regiments of British infantry charged into the village, capturing the baggage and many prisoners. The French cavalry fought their way out of the village, and forming up joined their infantry, which had formed square. Our artillery opened fire on the squares.

The 13th Light Dragoons captured the French artillery; and the regiment, commanded by Captain George Gore, together with the 2nd Hussars, King's German Legion, put the French cavalry to flight, after which the French force was completely overpowered. A French General, Brune, who tried to escape in a carriage, was overtaken by a few men of the regiment. He shot two dragoons, but finally gave up his sword to Trumpeter Martin.

Lieut.-General Rowland Hill, in his despatch recording the conduct of the troops employed in this engagement, states: "No praise of mine can do full

1811. justice to their admirable conduct: the patience and goodwill shown by all ranks during forced marches in the worst of weather, their strict attention to the orders they received, the precision with which they moved to the attack, and their obedience to command, during the action: in short, the manner every one has performed his duty, from the first commencement of the operation, merits my warmest thanks."

The regiment continued with the force under Lieut.-General Rowland Hill during the winter of 1811 and early part of 1812, but were not actively employed against the enemy.

1812. After the capture of Ciudad Rodrigo by Lord Wellington in January, the regiment formed part of the covering army advancing on Badajoz under that commander. In March they moved from Albuquerque upon the town of Merida, both situate in the Spanish province of Estre-Medura. Near Merida the regiment had a sharp encounter with the enemy, and Sergeant-Major Dunwoody took a French colonel prisoner.

From Merida the regiment, still with the force under Lieut.-General Rowland Hill, moved upon Almendralejos, to cover the troops besieging Badajoz, which place was taken during the night of the 6th April. After the capture of this fortress Lord Wellington had to alter his plans, and instead of invading Andalusia marched towards Beira, where, from the want of activity and firmness of the officer left in command on this frontier, matters were not satisfactory.

The regiment remained in Estre-Madura with the

force under Lieut.-General Rowland Hill, and was 1812.
engaged in a skirmish at Polomor on the 22nd May.

Towards the end of July, the regiment, with the brigade under Major-General Long, was engaged near Ribera; the gallantry of a detachment under Captain Benjamin Handley being specially mentioned.

After the decisive victory gained by Lord Wellington at Salamanca, July 22; Lieut.-General Hill's troops moved into the centre of Spain, and in September occupied a line from Toledo to Aranjuez, the cavalry going forward to Belmonte. In October, however, these troops were withdrawn and fell back on the main army at Salamanca, the regiment being engaged in a small skirmish on the 25th of the month near Ocana, as also in repeated skirmishes from the 9th to the 15th November near Alba de Tormes.

The regiment formed part of the rear-guard when the army retired to Ciudad Rodrigo, and was constantly engaged with the enemy. Many horses were lost from inclement weather and scarcity of forage.

In the spring, it was ordered to transfer its horses 1813.
to other regiments and return to England. It proceeded (dismounted) to Lisbon, embarked at Belem on the 17th April, and landed on the 17th May at Spithead, from whence it proceeded to Northampton.

On the 4th March, Lieut.-Colonel Charles Morland (who had been a major in the regiment in 1810, and purchased a lieutenant-colonelcy in the 50th Foot) exchanged back to the regiment with Lieut.-Colonel Chabot, and was appointed to the command. He assumed command

1813. on the 26th May, and by his great zeal and activity soon brought the regiment to a very high state of efficiency.

1814. The establishment of the regiment was reduced from ten to eight troops. In September it embarked at Bristol for Ireland, but was detained so long at sea by bad weather, that on landing at the Cove of Cork towards the end of December, the greater part of the horses were found to be lame from fever in the feet.

1815. The return of Bonaparte to France once again made war in Europe probable, and the regiment was recalled to England. It embarked at Dublin on the 4th May, landed at Liverpool on the 7th, and was quartered at Manchester.

In commemoration of its services in Portugal and Spain, it was honoured with the Royal Authority to bear the word "Peninsula" on its guidons and appointments.

In December, pantaloons and hussar boots were abolished, grey overalls and short boots being substituted.

An issue of Standing Orders for the regiment was published this year.

1816. In February, a detachment of the regiment under Captain John G. Peters, marched to London by order of H.R.H. the Commander-in-Chief, and was in the Riding School at Pimlico, together with detachments from the 12th, 16th, and 23rd Light Dragoons, instructed by that officer in the "Exercise of Lancers."

On the 27th July the regiment was reviewed on

Hounslow Heath by H.R.H. the Prince Regent, accompanied by H.R.H. the Duke of York, Commander-in-Chief, and on the 29th the following letter was sent to the Colonel of the regiment, General the Earl of Rosslyn:— 1816.

HORSE GUARDS, *29th July* 1816.

It is a great personal gratification to the Commander-in-Chief to have the opportunity of expressing to General the Earl of Rosslyn, and through His Lordship's means, to convey to Lieut.-Col. Morland, and to the officers, non-commissioned officers and privates of the 9th Light Dragoons, his entire approbation of the appearance of the Regiment when reviewed on the 27th inst.

The condition of the horses, the uniformity and excellence of the appointments, the horsemanship of the officers and men, and above all, the regularity and precision of their movements, indicate a general attention and exertion which reflect much credit on every individual belonging to the Regiment.

(Signed) RALPH DARLING,
Deputy Adjutant-General.

The experience of the war having proved most fully the importance of cavalry armed with the lance, on 19th September H.R.H. the Prince Regent was pleased to direct that the 9th, 12th, 16th, and 23rd Regiments of Light Dragoons be armed and equipped as "Lancers," and in consequence, discontinue to carry a carbine. From this date the regiment was designated the 9th Lancers.

A revised edition of the Standing Orders of the regiment was issued this year. 1819.

On the 24th July a brigade of Lancers, composed of the 9th, 12th, and 17th, was reviewed by H.R.H. the Prince Regent at Hampton Court.

1821. In July, the regiment was on duty at the Coronation of His Majesty King George IV., and on the 24th, being brigaded with the 2nd (or Royal North British) Dragoons and the 4th Light Dragoons, was reviewed by H.R.H. the Duke of York, Commander-in-Chief, on Wormwood Scrubs.

In August, the establishment was reduced from eight to six troops.

1822. On the 8th May, the regiment marched from Richmond to relieve the 10th Hussars at Brighton.

1823. In May and June, the regiment embarked at Bristol and Bideford, for Ireland; where it was quartered till March 1826.

1826. In March it embarked at Donaghadee, County Down, for Portpatrick; and Headquarters proceeded to Glasgow.

1827. In March the regiment moved from Scotland to England, and arrived at York the beginning of April.

1828. On the 4th March, Colonel Charles Morland, who had commanded for fifteen years—who for his devotion to and care of it, may well be termed the "Father Lieutenant-Colonel" of the regiment—was, on account of ill-health, obliged to retire on half-pay; being succeeded as Lieut.-Colonel Commanding, by Lord Loughborough on the 10th April.

The following letters from the Colonel, Lord

Rosslyn, and Lieut.-Colonel Lord Loughborough, are 1828. of interest :—

COPY LETTER from the Earl of Rosslyn to Colonel Charles Morland, 9th Lancers.

ST. JAMES' SQUARE, *6th April* 1828.

MY DEAR MORLAND,—I cannot leave town without expressing to you how deeply I regret the events that have induced you to retire from the command of the Regiment, and how much I lament the loss that the public service will sustain by your resignation of the command. I need hardly add how sensible I am of the degree in which I shall myself suffer by it, or of the advantages of which the Regiment will be deprived by your exchange. The high character which the Regiment has justly acquired for all its interior economy and discipline, as well as for its military instruction, sufficiently attest your merit as an officer, and the esteem and confidence with which all ranks and descriptions of those who have been under your command regard you are such as cannot fail to be highly flattering and satisfactory to yourself, and most gratifying to your friends.

I cannot conclude without expressing my hope that quiet and repose will restore your health in all its vigour, and repeating the assurance that I shall always have the greatest pleasure in bearing my testimony to the distinguished merit of your services, and in promoting, by any exertions in my power, any object which may appear to be desirable for you and suited to your situation.

With respect to the Regiment, I can only wish that it may continue in the same system in which you have established it, and lose nothing of the credit for which it is mainly indebted to your command.—Yours faithfully, ROSSLYN.

1828. COPY LETTER from Lord Loughborough to Colonel Morland, 9th Lancers.

LONDON, *26th March* 1828.

MY DEAR COLONEL,—You will have seen by the papers that our exchange is at last notified in the *Gazette* of last night. The pleasure which I naturally feel at finding myself at the head of the Regiment is not a little diminished by the regret which I entertain, in common with everyone who has had the pleasure of serving under your command, for the cause which has induced you to retire, and by a strong sense of distrust of my own abilities to keep the Regiment in that high state of discipline and good order to which your talents and exertions first raised, have since for so many years retained it. I shall not, however, scruple to avail myself of your kind permission to apply to you for the assistance of your advice and experience in all cases of difficulty that may occur, and I feel confident that in so doing, and endeavouring to adhere in my command of the Regiment as closely as possible to your system, you will consider that I am giving the best proof of the sincerity of my feelings and respect for you as a friend and commanding officer.

My joining the Regiment will not, I hope, be very long delayed, as my child is much better, and it will give me great pleasure if I can be of any use to you, either here or at York, in facilitating any arrangements you may wish made.

With best respects to Mrs. Morland from myself and Lady Loughborough, Believe me, my dear Colonel, Yours faithfully,

LOUGHBOROUGH.

To Col. Morland.

COPY LETTER from Lord Loughborough to Colonel Morland, 9th Lancers.

YORK BARRACKS, *13th April* 1828.

MY DEAR COLONEL,—Your late brother officers of the 9th Lancers, having unanimously resolved to request your acceptance of a piece of plate with which it is their wish to present

you as a mark of the esteem and affection which they feel for 1828.
you as a friend, and as a memorial of their respect for you as their Commanding Officer, have requested me to communicate to you their resolution, and to express to you their hope that you will not reject this testimony from a corps of officers who, having so long had the pleasure of serving under your command, feel the warmest interest in your happiness and welfare, and whose regret at your retirement from amongst them can only be diminished by the confident hope which they entertain, that it will be the means of restoring you to the enjoyment of perfect health, and enable you at a future period to return to the service in a higher situation.

Allow me to assure you that no one more deeply regrets the cause that induced your retirement, feels more interested for your happiness, or more sensible of your friendship and kindness, than myself.

With the strongest feelings of respect and esteem, Believe me, my dear Colonel, Yours faithfully,

LOUGHBOROUGH.

To Col. Morland.

CHAPTER IV

1829–1842

1829. In December, the colour of the overalls was changed from blue-grey to dark-grey, the lappels were abolished, and a black plume was substituted for a red-and-white one.

1830. In April the regiment entered upon the King's duty, and on the 26th June, the first escort to H.M. King William IV., (on his accession to the throne,) was furnished by it.

On the 14th July, a squadron of the regiment, under Major Alexander Campbell, marched to Slough to attend the funeral of H.M. King George IV.

On the 22nd, the regiment marched from Hounslow to Hampton Court under command of Lieut.-Colonel Alexander Campbell, for inspection by H.M. the King, who in the *London Gazette* of the 24th July, was pleased to direct that the regiment should assume the title of the 9th (Queen's Royal) Lancers, in honour of His Majesty's Royal Consort, Queen Adelaide.

By General Order of 2nd August, the wearing of moustachios was abolished except for the Life Guards, Horse Guards, and Hussars. During this year the regiment furnished many guards of honour and escorts to the King. 1830.

During the winter of this year it was detached to various places in Hampshire, Wiltshire, and Somersetshire, and assisted in keeping order among the agricultural population, who had resorted to incendiarism, and damage to property generally.

Early in this year the regiment returned to Hounslow, and in May and July two squadrons paraded at Windsor to attend the ceremony of presentation by King William IV., of a pair of silver kettledrums to the 1st, as also to the 2nd Regiment of Life Guards. 1831.

On the 8th September it attended the Coronation of their Majesties King William IV. and Queen Adelaide.

In January, in accordance with a General Order of August 1830, the clothing and appointments of the regiment were altered. 1832.

In March, one 2- and one 3-year-old cream-coloured colts (from the Royal Stud) were presented to the regiment by His Majesty, for the use of the band.

On the 21st April, it was inspected by His Majesty in the new clothing, who was pleased to express his entire approbation thereof.

Her Majesty Queen Adelaide, made a donation of £50 to the regiment's Benevolent Fund.

In June the regiment embarked at Bristol for Ireland, and proceeded to Dublin, where it was employed in aid of the civil power.

1832. A revised edition of the Regimental Standing Orders was issued this year.

1833. It was ordered that the sabretache should be worn when dismounted.

1835. On the 14th May the regiment embarked at Dublin for Liverpool, and on arrival in England was quartered at Coventry.

1836. In the spring of this year the regiment marched to Scotland, and arrived in Edinburgh in May.

On the 25th July, a sergeants' mess was first established by Regimental Order.

In December it was inspected by its Colonel, General the Earl of Rosslyn, G.C.B.

1837. On the 18th January, he died at Dysart House, Fifeshire, and was succeeded by Lieut.-General Samuel Need, who commenced his army career in the 1st Dragoon Guards in 1784.

In April, the sergeants' mess was abolished by order of Lieut.-Colonel the Earl of Rosslyn, who had succeeded to this title on the death of his father, as above mentioned. In April the regiment moved to Glasgow.

By Horse Guards letter, 12th August 1837, it was ordered that the regiment should retain upon its buttons and appointments the cypher W.R., instead of adopting that of V.R. as directed by Order 31st July 1837.

There is no official record until 1846, of the date or authority by which the regiment first bore its present cypher of "A.R." reversed. It appears probable however, that when it was authorised to assume the title of (Queen's Royal) Lancers, it would at the same time (viz. 1830) be permitted to bear the Queen's cypher of "Adelaide Regina," and W.R. must have been a clerical error for A.R. In February 1833, a silver medal was presented to Regimental-Sergeant-Major Lindsay on leaving the regiment, and on it was engraved our present cypher. 1837.

In May, the regiment marched from Scotland to England, and was quartered at Leeds, moving in the autumn to Nottingham. 1838.

In April, a portion of the regiment was inspected by the Colonel, Lieut.-General Samuel Need. He died on the 18th August, at Fountain Dale, Notts, and was succeeded on the 24th August by Major-General James Wallace Sleigh, C.B. (who commanded the 4th Cavalry Brigade at the battle of Waterloo). This officer inspected the regiment in November at Ipswich, to which place it had moved in April. 1839.

The Colonel, Major-General James W. Sleigh, on the 14th March directed that officers and men should discontinue to shave the upper lip, which had already been ordered in August 1830. 1840.

In April the regiment marched to Hounslow, and took up travelling escort duty for Her Majesty Queen Victoria.

By Horse Guards letter, 20th May 1840, it was ordered that in consequence of the alteration in the

1840. colour of the clothing from red to blue, the facings should be scarlet instead of blue.

In October, the regiment under command of Lieut.-Colonel Alexander Campbell, K.H.,[1] attended the funeral of H.R.H. the Princess Augusta, sister of the late King (William IV.)

1841. In April it marched from Hounslow to Dorchester, being relieved at the former place by the 11th Hussars.

1842. In January, the Earl of Rosslyn, who as lieutenant-colonel had commanded the regiment for fourteen years, retired from the army, being succeeded by Lieut.-Colonel Alexander Campbell.

In February, Messrs. Hawkes and Masky were appointed regimental cap and accoutrement makers.

[1] "K.H.," Knight of the Hanoverian Guelphic Order.

CHAPTER V

1842–1853

In April the regiment was placed under orders for India, and in May was inspected by the Colonel, Lieut.-General J. W. Sleigh, C.B., who complimented it highly on its smart appearance, on its high state of discipline and efficiency, and who praised most strongly the fitness and condition of the horses (258), which were handed over to the 4th Light Dragoons at Canterbury. 1842.

Between the 7th and 31st May, the regiment embarked in five transports at Gravesend for Calcutta, where it arrived between the 25th August and 23rd September, and was quartered at Fort William *en route* for Cawnpore, which was reached in January 1843.

On the 1st May 1842, the strength of the regiment was 743 all ranks, being only one below the establishment. In June of this year it was 90 short, this number having died of cholera and fever coming from Calcutta to Allahabad by boat—viz. 1 officer, 5 sergeants, 2 trumpeters, 83 rank and file—as also 14 women, and 8 children. 1843.

On the establishment of a regency in the Gwalior State internal disturbances arose, and the British had

1843. to intervene. A portion of the regiment (Captain Read's, Captain Rose's, Captain Macartney's, and Captain Power's troops) joined the left wing of the army of Gwalior, under General T. Grey, C.B., on the 25th November; and was engaged in the rout of the rebels at Punniar on the 29th December. Another action, under Sir Hugh Gough, Commander-in-Chief, took place on the same day at Maharajpur, at which Colonel John Scott of the regiment commanded a cavalry brigade.

1844. On the 24th January, that portion of the regiment which had been engaged with the Gwalior force returned to Cawnpore. Lieut.-Colonel Campbell, K.H., was appointed a Companion of the Bath.

By General Order in India, 21st August, the regiment was permitted to wear on its colours and appointments the word "Punniar"; and a bronze star made from the guns captured at this place was presented by the Government of India to the officers and men of the Royal Service who were engaged in that action.

1845. In October, the regiment moved to Meerut, but in December was ordered to join the army of the Sutlej, under General Sir Hugh Gough, G.C.B.

Colonels Campbell and Scott being in command of the 1st and 2nd Cavalry Brigades, the command of the regiment devolved upon Brevet-Lieut.-Colonel J. Fullerton, K.H. It was formed in brigade with two native infantry regiments, two 12-pounder elephant batteries, and a company of sappers; the whole under Lieut.-Colonel Fullerton. The cavalry joined Sir Hugh Gough near Sutlej, fifteen miles from Ferozepore.

On the 10th February the regiment took part in the battle of Sabraon, which was a sequel to that of Ferozeshah, and completed the first phase of the Sikh war. Sabraon, being an attack on an intrenched position, was more an artillery and infantry battle, than a cavalry action. 1846.

A squadron of the regiment under Hope Grant was sent forward to make the enemy reveal his guns, which duty was most effectively carried out. Campbell's brigade, consisting of the regiment and two troops Horse Artillery, was on the right flank, in the rear between General Gilbert's and Sir Harry Smith's Divisions. Scott's Brigade, consisting of the 3rd Light Dragoons, 3rd and 9th Irregular Cavalry, and also the 4th, 5th, and 73rd regiments Native Infantry, was on the extreme left, in rear of Sir Robert Dick's Division.

The regiment formed, with the Body-Guard and a troop of Horse Artillery, the Governor-General's escort for the great Durbar at Lahore, held in the City Palace of the Maharajah.

The Sikh rebellion being for the time quelled, the regiment returned in March to Meerut. A medal was granted to officers and men of Her Majesty's and the East India Company's armies who were engaged at any of the following battles — Moodkee, Ferozeshah, Aliwall, and Sabraon.

In June of this year, authority was given for the word "Sabraon" to be borne on the colours and appointments. 1847.

The following incident which happened to Sir Hope

1847. Grant will be read with interest:—Hope Grant had a narrow escape of being shot on a court-martial parade. An artilleryman had been sentenced by general court-martial to be shot for striking his superior officer. A volley was fired, and the provost-sergeant, who was in charge of the firing-party, thinking it had not taken effect, went forward and fired a pistol close to the head of the prisoner, who had been posted for the execution of the sentence within the flank of the regiment. The bullet went through the prisoner's head, whistled past Grant, and went through the cap of a soldier standing behind him.

1848. On the 29th September, the regiment moved from Meerut and joined the army of the Punjab under Lord Gough, G.C.B.

On the 4th December it, with the 14th Light Dragoons and Horse Artillery, under Sir Walter Gilbert, crossed the river Chenab from the left to the right bank to pursue the Sikhs, who had fallen back before the force under General Thackwell. On the 5th they arrived at Heylah, from whence a reconnaissance under Major Hope Grant was made, when finding the enemy still retreating, a return was made to Heylah, where the cavalry remained until the 11th January 1849.

1849. On the 13th January, Lord Gough decided to attack the enemy who were in position to the north of Chillianwallah.

An exhaustive account of this engagement forms no part of this record. The regiment, under command of Lieut.-Colonel Fullerton, unfortunately formed part of the cavalry brigade on the right flank of the attacking line, being brigaded with the 14th Light Dragoons and the 1st and 6th regiments of Light Cavalry, under Brigadier Lieut.-Colonel Pope of the 6th Light Cavalry. Through some mistake, the brigade advanced and covered the fire of Major Christie's battery of Horse Artillery. Before this could be rectified a body of the enemy's cavalry charged down on the then stationary brigade, and undoubtedly a panic ensued. Guns and horses were captured, and two historians, Thackwell ('The Second Sikh War') and Malleson ('Decisive Battles of India'), relate that but for the portion of the 9th, then engaged, having rallied in rear of the guns and stayed the rush of the enemy, the Commander-in-Chief and his staff would have been assailed. The following is Lieut.-Colonel Hope Grant's official account of the incident :— 1849.

CHILLIANWALLA CAMP, 15*th Jan.* 1849.

SIR,—I have the honour to report, for the information of Major-General Sir Joseph Thackwell, K.C.B., that two squadrons of the 9th Lancers, in the action of the 13th inst., were in line on the extreme right of the four regiments constituting the 3rd brigade of cavalry.

The two squadrons which I took command of, on the right of the 2nd brigade, were proceeding on steadily, and changing their direction a little to the left, when the native cavalry began to cheer and charge. I confess at the time I could see no enemy, except a party of about fifty horsemen a good deal to our right flank, which, from having red coats on, in the distance I took to be some of our own Irregular Horse, as they were apparently going in the same direction as ourselves.

The 9th Lancers were dressing upon the 6th Light Cavalry I think, and they of course increased their pace, and brought their lances down to the guard. There were some few of the

1849. enemy now seen in our front, but nothing in the force to stop any body of Europeans.

I had an encounter with one of them, and received two blows with the sword, but not sufficient to go through more than the "murzir" which I wore.

The two squadrons were going along with the line steadily, and no hesitation was evinced; on the contrary, the flank men were engaged with some of the enemy, and doing their duty, when the whole line checked and went about from the left, and my squadron, certainly without a word from me, turned round too; but the jungle and dust might make some excuse for the men, as it was difficult to hear, and in many cases to see.

The dust from this movement became very great, and the men of my regiment got mixed up with the other regiments; and though I did all in my power to stop them, ordering them to halt and front, and many of the officers in the regiments did the same, it was useless: they would not turn round.

They appeared after having gone about to have got panic-struck. I grieve greatly to make this report of my regiment, as I did consider them, and even do so still, as fine a set of soldiers as there are in the army; but it was one of those wonderful coincidences which there is no accounting for. I grieve too for myself, as it must ruin my prospects in life.—I have the honour to be, &c., &c.,

J. HOPE GRANT, Lt.-Col.

To the ASSISTANT ADJUTANT-GENERAL,
Cavalry Brigade.

The other two squadrons of the regiment, in command of the Honourable Charles Powys and Major Yule, with two of the 6th Light Cavalry, had under Colonel Lane, Horse Artillery, repelled the attack of a large body on the extreme right. The engagement appears to have been a most unsatisfactory and undecided one from very many points of view.

During the ensuing three weeks the two armies watched one another, until on the 20th February Lord Gough decided that he was able to again give battle to the Sikhs, and this took place at Goojerat on the 21st February. 1849.

The regiment, Lieut.-Colonel Hope Grant in command, together with the 3rd Dragoon Guards, 8th Light Cavalry, and the Scinde Horse, were under command of Brigadier White on the left flank of the line, supported by Duncan and Huish's troops of Horse Artillery. Large bodies of the enemy's cavalry threatened this flank, and General Sir J. Thackwell, who was in command of this portion of the line, ordered a squadron of the 9th, and the Scinde Horse, supported by another squadron of the regiment under Hope Grant, to charge. They drove everything before them and captured two standards. Various squadrons of the regiment took part in the final chase and defeat of the Sikh army.

The regiment moved into quarters at Wuzeerabad, and lost by sickness seventy men this year.

By Horse Guards Order, 27th November, the stripe on the overalls was changed from scarlet cloth to yellow lace.

Colonel A. Campbell having died of quinsy in March, and the next senior, Lieut.-Colonel Fullerton, C.B., also dying while on leave in Cashmere from over-exertion while crossing a pass, Lieut.-Colonel J. Hope Grant was on the 29th April appointed to the command of the regiment. With reference to Lieut.-Colonel Fullerton's death, Lieut.-Colonel H. Marshall in 1875 1850.

1850. found his grave in the graveyard at Sirinigar, with a tombstone over it erected by his brother officers, on which it was recorded that he died at Ramoo, 28th April 1850.

In December, Major Little who was in temporary command, applied for twelve carbines per troop to be issued to the regiment. He made this application on the suggestion of Majors Pratt and Yule, who had served with the 16th Lancers in the Afghan (1842), Gwalior, and Sutlej campaigns, when that regiment was armed with the carbine in the proportion above named, and had found it a most useful weapon—advance-guards, flanking, patrolling, and skirmishing parties being always armed with it. The application was also supported by General Sir Walter Gilbert, who had commented somewhat severely on the want of such a weapon for Lancers.

Despite these recommendations, Major Little was severely snubbed by the Inspector-General of Cavalry at home (Major-General Brotherton), and nothing came of the suggestion. As we see later, 1878 produced the carbine, and 1903 a rifle, for Lancers.

1851. On the 5th February, having marched from Wazeerabad, the regiment reached Umballa, and was there quartered.

1853. By General Order in India, dated 9th March, the regiment was permitted to wear on its colours and appointments the words "Chillianwallah" and "Goojerat."

CHAPTER VI.

1857–1859.

Consequent on the outbreak of the Mutiny a force was formed at Umballa, called the Delhi Field Force, under the supreme command of General Anson, Commander-in-Chief in India. 1857.

The formation of it was as follows :—

1st Umballa Brigade: Brigadier-Colonel Halifax, 75th Regiment.	75th Regiment (Queen's). 1st Bengal Europeans. Two squadrons 9th Lancers. One troop Horse Artillery (Turner's or Money's).
2nd Umballa Brigade: Brigadier-Colonel Jones, 60th Royal Rifles.	2nd Bengal Europeans. 60th Native Infantry. Two squadrons 9th Lancers. One squadron 4th Bengal Lancers. One troop Horse Artillery (Money's or Turner's).
Meerut Brigade: Brigadier-Colonel A. Wilson, Royal Artillery.	Wing 60th Royal Rifles. Two squadrons Carabiniers. One field battery (Scott's). One troop Horse Artillery (Tomb's). Native sappers. 120 siege artillerymen.

1857. DISTRIBUTION OF THE OFFICERS OF THE REGIMENT, MAY 1857.

Rank.	Name.	Present or Absent.	Remarks.
Lieut.-Colonel	J. Hope Grant, C.B.	Present	Cavalry Brigadier.
,,	A. Little	Leave, England	Rejoined 28th Oct. 1857.
Major	J. R. Rose	,,	Promoted to 17th Lancers, 18th Sept. 1857.
,,	H. A. Ouvry	Leave, Cashmere	Rejoined 25th June 1857.
Captain	The Hon. C. Powys	Leave, England	Rejoined 28th Oct. 1857.
Brevet-Lieut.-Colonel	R. Yule	Leave, Cashmere	Rejoined 6th June 1857; killed on 19th June.
Captain	W. Drysdale	,,	Rejoined 24th June 1857.
,,	O. St. G. Anson	Present	Temporary command.
,,	A. F. Steele	Leave, England	Rejoined 15th Nov. 1857.
,,	J. R. Coles	,,	Rejoined 28th Oct. 1857.
,,	C. P. Johnson	,,	Rejoined 12th Nov. 1857.
,,	A. M. Fawcett	Present.	
,,	T. W. Hutchinson	,,	
,,	H. Hamilton	,,	Brigade-Major.
Lieut.	L. J. French	Leave, Cashmere	Rejoined 24th July 1857.
Brevet-Captain	J. Head	Present.	
Lieut.	H. A. Sarel	Leave, to hills	Rejoined 24th June 1857.
,,	F. Ellis	England	Depôt troop.
,,	A. E. Wilkinson	Leave, Cashmere	Rejoined 10th Aug. 1857.
,,	H. Scott	Staff employ, Bombay	Rejoined 3rd Feb. 1858.
,,	R. Shaw	Present.	
,,	A. E. Willis	Leave, England	Rejoined 17th Dec. 1857.
,,	C. D. Rich	Depôt, Maidstone	Rejoined 28th Oct. 1857.
,,	R. Mills	Leave, England.	
,,	R. D. Upton	Present.	
,,	A. S. Jones	,,	
,,	J. Goldie	,,	
,,	A. Bell-Martin	,,	
,,	E. King	Depôt, Maidstone.	
,,	J. de H. Chadwick	Leave, England	Rejoined 17th Dec. 1857.
Cornet	J. Evans	Present.	
,,	A. Morrogh	Depôt, Maidstone	Joined 17th Dec. 1857.
,,	P. Thursby	,,	,,

DISTRIBUTION OF THE OFFICERS OF THE REGIMENT, MAY 1857—*continued.* 1857.

Rank.	Name.	Present or Absent.	Remarks.
Cornet	S. A. Pretor	In England	Joined 18th Sept. 1858.
Paymaster	T. H. Radcliffe	Leave, England.	
Adjutant	F. R. Grant	Present.	
Surgeon	H. Franklin	Leave, England.	
Assistant-Surgeon	J. Clifford	Present.	
,,	D. S. Smith	,,	
Veterinary Surgeon	R. J. Hurford	Leave to hills	Rejoined 8th Aug. 1857.
Quartermaster	P. House	Present.	
	Attached.		
Lieut.	R. Blair.		

Colonel Hope Grant was appointed Cavalry Brigadier of the force, with the Adjutant of the regiment (W. Hamilton), as his brigade-major; the command of the regiment devolved temporarily on Captain O. Anson.

The strength was as follows:—16 officers, including Grant and Hamilton; 493 N.C.O. and men. Ninety-five men were on furlough in the hills, and many officers away on leave in England and India.

The force moved in four divisions, a squadron of the regiment with each, the first leaving on the 17th May, the last on the 25th: with the latter was the Commander-in-Chief, who died of cholera at Kurnal on the 27th May, and Sir Henry Barnard succeeded to the command. It reached Alipore, about twelve miles north of Delhi on the 5th June, and was joined on the 7th by the Meerut brigade, when it was found that the rebels were in a strong position at a place called Budlee-ki-Serai, about five miles from Delhi.

At 1 A.M. on the 8th the force moved out to the attack. Showers, in command of the 1st brigade,

1857. attacked on the right; Graves, with the 2nd on the left; three squadrons of the regiment, and ten guns under Hope Grant, got behind the enemy's position, and when shaken by the artillery and infantry, they fell on their flank. Captain and Brevet-Lieut.-Colonel Yule, who had rejoined from leave in Cashmere two days previously, was in command of the regiment. The plan of operations worked all smoothly, and the enemy were obliged to retreat, which at first they did in an orderly manner; but the three squadrons of the 9th fell upon them so fiercely that the retreat became a rout, and they left all their stores, baggage, and guns behind them.

Lieut. Alfred S. Jones in this action, assisted by Lieut.-Colonel Yule with a few men, captured a 9-pounder from the enemy, sabred the gunners, and turned it on a village where the rebels were seeking shelter. Lieut. Jones for his gallantry on this occasion was awarded the V.C. Cornet John (Bashi) Evans had a narrow escape; he was in command of Sir Henry Barnard's personal escort (one sergeant, fifteen men, 9th Lancers). A round-shot came among them, which killed the Adjutant-General of the army (Colonel Chester), and Captain Russell, who were on Evans's right and left respectively. Lieut. Roger Upton had his horse shot under him, and both the doctors lost their horses. The regiment's loss at Budlee-ki-Serai was—killed, 1 sergeant, 2 corporals, 11 men, 20 horses; wounded, 11 men, 4 horses.

General Barnard saw that now was the time to make an impression on the rebels, so decided to drive them back, right into Delhi. Our troops advanced in

two divisions—one under Wilson towards the Subzee 1857.
Mundee, the other under himself towards the British cantonments: they swept their way on, and aided by a dexterous flank movement, the Ridge was ours. The day's fighting deprived the enemy of their shelter outside the walls of Delhi, and gave us the base from which the operations connected with the siege and capture of Delhi were conducted. The behaviour of the Sirmoor or Goorkha battalion under Major Reid on this, and many further occasions evoked much praise.

Colonel Hope Grant took as his A.D.C. the Hon. Augustus Anson, nephew of the late Commander-in-Chief.

On the 15th June, the respective squadrons were officered as follows:—

1st . . .	Lieut. R. Upton, Lieut. A. G. Bell-Martin.
2nd . . .	Lieut. J. Head, Lieut. R. Blair (attached).
3rd . . .	Captain Anson, Lieut. A. M. Fawcett, and Captain T. W. Hutchinson.
4th . . .	Lieut. A. Jones, Lieut. J. Goldie.

On the 17th June, Regimental-Sergeant-Major R. F. Thonger was given his Commission as cornet in the regiment. Sergeant Allan Deane was appointed regimental-sergeant-major in his place.

Late in the afternoon of the 19th, the rebels make an attack on the rear of the camp. Hope Grant, with a squadron of the regiment, two guns of Scott's, two of Turner's, and two of Money's batteries, turned out at once; the guns were in great peril, when Daly, with a detachment of the Guides' cavalry, charged, and for the moment saved the situation. The camp was

1857. soon aroused, and Grant was quickly followed by Lieut.-Colonel Yule with the other three squadrons, who charged a large body of the enemy on the Subzee Mundee Road. Daly was wounded in the shoulder when he made his first charge, and Lieut.-Colonel Yule was killed, in what manner has never transpired. His horse galloping into camp was the first intimation that he was missing. He is buried below the Ridge, and by his side Captain W. C. Clayton, who died at Delhi from the results of an accident at polo, December 1876.

As long as daylight lasted the rebels were driven back, but when darkness set in they got round our flanks; and two of our guns (Money's probably) were in jeopardy. Grant collected a few men and charged the rebels; a Sepoy five yards from him shot his horse, and he was left dismounted in the middle of the enemy. Three men stuck to him—his native orderly, of the 4th Irregulars, Rooper Khan; Privates Thomas Hancock and James Purcell, of the regiment. Purcell had his horse killed the same time as Hope Grant. Hancock and Rooper Khan both tried to induce Hope Grant to take their horses, he eventually took hold of Rooper Khan's horse's tail and was dragged out of the crowd. Purcell and Hancock were awarded the V.C. (Hancock was appointed a gate-keeper in Windsor Park, Purcell was killed before Delhi, September 14, 1857), Rooper Khan was promoted by his commanding officer, Major Martin, and Sir H. Barnard awarded him the second class of the Order of Merit.

Sir Hope describes this as the hardest day's work during the Mutiny. He did not return to camp till

11.30 P.M., and at 3.30 A.M. on the 20th, was out with 1857.
a force to try and pick up any guns, ammunition, &c., which might have been left on the field. He secured a gun and three ammunition-waggons.

20th June.—The following orders for stores by the mess president at this date may be of interest:—

Per Month.		
150 dozen	. . .	Soda-water.
50 „	. . .	Beer.
12 „	. . .	Claret.
9 „	. . .	Brandy.

At 2.30 A.M. on the 5th July a squadron of the regiment, one of the Carabiniers, 12 guns, and 1000 infantry, under Colonel Coke (Punjab Irregular Cavalry), went to cut off a party of rebels returning from Alipore; where they had gone to try and intercept a convoy of our wounded men on the way to Umballa, and another with ammunition and treasure coming from Ferozepore, as also to drive out a troop of the 5th Irregular Sikh Cavalry under Lieut. Younghusband; which was stationed at Alipore, to keep up our communications with Kurnaul, &c. The action was indecisive: the enemy however drew off, leaving his plunder and some ammunition behind.

General Sir H. Barnard died of cholera, and was succeeded by General Reid.

On the 9th July, a battery of heavy guns was menaced by the rebels, who by treachery had been let into the British camp. The battery was supported

1857. by a picquet (thirty men) of the regiment under Lieut. Bell - Martin, which coming to the rescue, saved the guns. Captain Light wrote thus of the conduct of the picquet: "The bold front and steadiness of the troop of the 9th Lancers under Lieut. Martin saved my guns. Nothing could be steadier."

On the 17th July, Brigadier-General Wilson, Bengal Artillery, succeeded General Reid, who had to resign through ill-health.

Thermometer at 5.30 P.M., 99°.

When the 4th Irregular Cavalry were disarmed; Hope Grant's two orderlies, Rooper Khan and Peer Khan, begged not to have to lay down their swords on the general heap. Hope Grant took them, and next day returned them, also allowing them to keep their horses; both men kept loyal, and in 1859 they visited Hope Grant at Lucknow, where he gave each of them a new sword.

In July, Major H. A. Ouvry assumed command of the regiment. He had a narrow escape while bathing at the Flag-Staff; a shell dropped close to him, so he fell flat on the ground under a bank, and while in that position a splinter came right over him.

On the 12th August, a column under Brigadier Showers, consisting of 1200 infantry, a troop of 9-pounders, the second squadron of the regiment under Anson, and the Guides' cavalry; left camp at 2.45 A.M. and surprised a strong picquet of the enemy close to Ludlow Castle, capturing four guns. Loss—19 men killed, 94 wounded.

The cavalry were not engaged, but had to sit under fire the whole time.

An officer at this time records his mess bill as under: Wine bill, June, Rs. 73, 14 ann. Messing, July, Rs. 70, 15 ann. 1857.

On the 10th August, Lieut. A. Jones, V.C., was appointed a Deputy-Assistant-Quartermaster-General.

A quaint form of sport was indulged in by Bashi Evans. On the 18th, he was for two hours up to his breast in water under a burning sun, surrounded by alligators, which he tried to catch by their noses.

On 25th August, the third squadron moved with other details, under Brigadier-General Nicholson, towards Nugufghur, where the enemy was reported to be in force, hoping to cut off our siege-train coming from Ferozepore. The ground was very wet and heavy, and progress slow. On the afternoon of the 26th, the force came up with the rebels and routed them utterly, capturing 13 guns, the standing camp, and quantities of ammunition. 800 were killed.

Lieuts. French, Blair, Sarel, and Goldie, and Cornet Evans, volunteered for service with the artillery, who were very short of officers and men. The services of Lieut. Blair and Cornet Evans were accepted, together with that of sixty men of the regiment; ten of whom did permanent duty in Major Turner's troop of Horse Artillery. The following letter from the officer commanding the Artillery Brigade is of interest:—

No. 23. *September 22nd*, 1857.

SIR,—The siege operations against Delhi having now terminated, and the detachments from H.M. 6th Dragoon Guards and 9th Lancers (10 men) which did duty with the artillery in the trenches having rejoined their regiments, it affords me the highest gratification in bringing to your notice, with a view to

1857. the same being communicated to the officers commanding those regiments, the excellent conduct of the men of those detachments, during the whole period they were attached to the artillery, and the valuable services they rendered. Every officer commanding a battery has reported most favourably of the men of the above detachments—their steadiness of conduct, their alacrity, and earnest desire to render themselves useful. In No. 2 battery especially, the officer commanding reports that on the morning of the 14th (Sept.), from a paucity of artillerymen, the guns were worked entirely by men of H.M. 9th Lancers. I desire also to bring to your notice the assistance rendered by Captain the Hon. A. Anson, in No. 2 battery, and to tender to that officer my best thanks for the services he rendered, and which I have not failed to report to the Major-General commanding the Forces.—I have the honour to be, Sir, Your most obedient servant,

J. GAIHKELL, Major,
Commanding Artillery Brigade.

To the Officer Commanding Cavalry Brigade,
Delhi.

Major Brind, who commanded the artillery during the siege, also wrote as follows: "I consider it incumbent upon me to bring to the special notice of Major-General A. Wilson the officers and men of the 9th Lancers who were employed in the batteries as artillerymen under my command and observation. Lieuts. Evans and Blair are deserving of honourable mention for the zealous and cheerful performance of their duties. The coolness and bravery of officers and men in situations of extreme danger whilst engaged in a novel duty, called forth the admiration of all who witnessed it."

On the 14th September, the final assault on Delhi took place. When the assaulting parties were drawn up near Ludlow Castle ready for the struggle, it was discovered that the enemy had during the night 1857. repaired the breach in the curtain of the Cashmere Gate, it was therefore necessary to open fire again from the 24-pounder battery. Only two sergeants of artillery were available to work it, so it was manned by thirty men from the regiment, who performed the work so well that Captain E. B. Johnson, Assistant-Adjutant-General of Artillery, wrote as follows to Sir Hope Grant: "Through the zeal and intelligence of your men, who worked like trained artillerymen, I was enabled to accomplish this (the re-establishment of the breach) with perfect success, and Brigadier-General Wilson personally thanked them when he came down to the battery for the very admirable service they had performed."

Two hundred of the regiment, with four hundred Sikhs, were sent under Grant for the protection and support of Reid's column, which was to enter Delhi by the Lahore Gate. Grant formed up his force under the city walls, somewhere close to the Moree Bastion; covering the whole of our batteries which had hitherto been unprotected, with the exception of that in front of Hindoo Rao's house, where there was a small force of infantry; as also keeping in check a body of about 5000 rebels with artillery, who were threatening our right from Kissengunge. When the Moree Bastion was taken, a few men of the regiment who had learnt how to work guns, dismounted; scrambled up the breach, and turned the abandoned guns on the enemy with great effect.

The attack on the Lahore Bastion having failed and left the defenders at liberty, Grant had to hold fast

1857. to his ground to protect the flank; when his force suffered most severely from a 24-pounder which was turned on it, as also from musketry-fire.

The loss of the regiment was 42 men and 61 horses killed and wounded; that of the Sikh cavalry, who were in support, 15 men and 19 horses.

The regiment was under the command of Captain Drysdale, whose forage-cap was hit with a bullet, and he and ten other officers of the regiment had their horses shot under them. Hope Grant was hit by a large grape-shot, which luckily glanced off and only bruised him severely. Lieut. Frank Grant was hit in the ribs in the same way. Captain Rosser of the Carabiniers, who was acting A.D.C. to Grant, was hit in the head with a bullet and hurled to the ground. Grant's orderly, Peer Khan, 4th Irregulars, had one horse killed and two wounded. Kaye, in his work, 'A History of the Sepoy War in India,' writes thus of the conduct of the regiment on this occasion: "The courage of the 9th Lancers was never surpassed: when Grant commended their sturdy gallantry under fire, they replied that they were ready to take as much more of it as he liked." For two long hours the brigade stood firm as a rock.

On the 20th September the city of Delhi was finally captured.

On the 24th the regiment, with Remmington's and Blunt's batteries of Horse Artillery, Major Bourchier's battery, and detachments of the 1st, 4th, and 5th Punjab Cavalry and Hodson's Horse (these latter totalling 400) under the command of Major Ouvry of the 9th, together with infantry; moved with Colonel Greathead's force from Delhi, to scour the Gangetic Doab for the fugitive rebels. 1857.

At an early hour on the 28th, the column was within about 1½ miles of Boolundshuhur, where the rebels made a stand. The regiment here did very good work, but suffered somewhat severely. Captain Drysdale, who was in command of the regiment, had his horse shot under him while going at full speed, and in the consequent fall broke his collar-bone. Lieut. Sarel had the forefinger of his right hand shot off as he was running a Sepoy through with his sword, and Cornet Thonger was also wounded. Lieut. Blair, who had been sent with ten men of the regiment to bring in an abandoned ammunition-waggon, was surrounded by a party of about 50 sowars; they forced their way through, but Blair, while in the act of running a man through, was cut down by a rebel of the 14th Irregular Cavalry, the blow falling on the top of his shoulder and severing the joint. The arm was taken out of the socket, a piece removed, and the limb then replaced. He was awarded the V.C. for his gallantry, as also was Private Donohoe, who helped Blair after he was disabled. Drysdale was recommended by Lieut.-Colonel Ouvry for the V.C. Trumpeter Robert Kells and Private Henry Jordan were awarded the V.C. for rescuing Captain Drysdale when he was disabled as above mentioned. Jordan died soon afterwards from a bad musket-ball wound. Private James Reynolds Roberts also earned the V.C. for having rescued Farrier Stillman, who in a charge through the town of Boolundshuhur, fell wounded, unobserved by all but Roberts, himself badly wounded in his bridle hand.

1857. The action lasted from 6 A.M. to 9 A.M. The enemy lost three guns and a quantity of baggage and ammunition.

On the 6th October, the regiment was employed chasing some rebels from Allyghur.

On the 7th, a curious accident happened. A man (Evans, G troop) and horse fell down a well about 14 feet deep. Luckily the horse went down hind-legs first, and they were hauled up with ropes.

On the 9th the force was at Bijegarh, about forty-eight miles from Agra. In consequence of urgent request for help from the Chief Commissioner, the cavalry and horse artillery were sent on at midnight, Greathead following four hours later with the infantry. On the morning of the 10th his force was under the walls of Agra.

Greathead having been given to understand by the Agra people, that on hearing of his approach the enemy had withdrawn over the Kallee Nuddee, a stream about ten miles distant, prepared to encamp his force on the parade-ground without first placing picquets and sending out patrols. The men were at breakfast, pitching tents, &c., those of Evans's squadron were changing their clothing. Suddenly the enemy appeared, their cavalry making a dash on our guns. This was gallantly repulsed by Captain L. J. French's squadron of the regiment, which drove them back in great disorder. French was killed, and Lieut. Jones was struck off his horse by a shot, and when on the ground was fearfully hacked about the head (he lost an eye), back, and stomach—twenty-two wounds in all. Our infantry and guns coming up, the rebels gave way, and were pursued by the cavalry and horse artillery beyond the river, absolutely beaten. Thirteen guns and much ammunition were brought into camp. 1857.

This had been a hard day's work. From first to last the cavalry and artillery had marched sixty-four miles, the infantry fifty-four, in less than thirty-six hours, besides fighting a general action. It was 7 P.M. before the force returned to camp.

Prior to the general assault, four natives, disguised as musicians or conjurers, had strolled up to the advance-guard of the regiment. The non-commissioned officer in command (Sergeant Crews) ordered them away, when one drew a concealed tulwar and killed him. Sergeant Hartigan, who was by, rushed up, and though he received a severe sword-cut on the head, wrenched the weapon away from the man and killed him with it, wounding another also. The other two were soon dispatched by the guard. Sergeant Hartigan was awarded the V.C.

When the regiment left Delhi, Hope Grant had to remain there, much to his regret at having to part from his own officers and men. One morning, when thinking over his sad fate, he received a letter from Mr Muir (then Secretary to the Government of the N.W. Provinces), dated Agra, 10th October 1857, in which the following passages occurred: "You are to come on as sharp as you can"; "You are to come on at once, in the mail-cart if possible." Grant got this letter countersigned by General Penny (who was then in command at Delhi in place of Wilson, who had gone on sick-leave to Mussoorie), with an order to assume command of the column (which had left Agra for

1857. Cawnpore) vice Greathead. On the 18th October he overtook it at Firozabad, and on the 19th, the fort of Mainpuri was taken without a shot being fired.

On the 23rd, Ouvry who was in command of the advance-guard, heard that 500 of the rebels and four guns with ammunition, had only left a place called Kanuj two hours in front of our troops. Grant sent him off with the first squadron of the regiment under Fawcett, two squadrons of the Punjab Cavalry, and two guns. Hearing sharp firing, Grant set off with another squadron and four more guns. On the bank of the Kallee Nuddee he found a party in charge of the enemy's four guns, which had been abandoned; the rest of our force having crossed the river in pursuit of the enemy, who were in flight towards the Ganges. One of the rebels, pursued by a 9th Lancer, dismounted and plunged into the river, the Lancer ordered him to stop, and the man came out of the river and walked towards him, he pointed his pistol at his breast and fired, with no result. The man then ran to the water again and dived, and on coming up was fired at again and again (Major Ouvry states he thinks a thousand shots were fired), but he succeeded in escaping (well deserved).

On the 26th the regiment reached Cawnpore, and on the 28th, Lieut.-Colonel Little, Captains Powys and Coles, and Lieut. Rich joined from leave in England, Lieut.-Colonel Little assuming command.

On the 30th, Hope Grant's force crossed the Ganges on the way to Lucknow, and halted at Bunteerah for the coming of the Commander-in-Chief (Sir Colin Campbell), driving the rebels out of the place and capturing a gun; three squadrons of the regiment being engaged. 1857.

The fourth squadron, which had been left at Cawnpore as escort for the Commander-in-Chief, marched from there to Bunteerah, forty miles in the day, on the 9th November. One day, when on convoy duty, a curious incident occurred. In passing through a clump of trees Lieut. Evans, with the advance-party, saw a nest of bees hanging from a branch, and thoughtlessly ran the point of his spear (most of the officers of the regiment carried a hog-spear at this time) through what looked like a brown-paper bag. The bees rushed out and put the advance-party to rout. Colonel Adrian Hope who was in command of the convoy, formed up his men to resist as he thought, an attack of the rebels.

On the 11th, the whole relieving force was inspected by Sir Colin Campbell, and Malleson, in his 'History of the Indian Mutiny,' thus describes the regiment: "The 9th Lancers, with their blue uniforms and white turbans twisted round their forage-caps, their flagless lances, lean but hardy horses, and gallant bearing, looked the perfection of a cavalry regiment on active service."

The regiment was employed during the relief of Lucknow in protecting the rear of the column and keeping open the line of communication with Dilkoosha, no event worthy of record occurring. A curious accident happened to Lieut. Frank Grant's charger. He had ridden into the Secundra Bagh, and left his horse with an orderly near a small building which was smouldering: some powder must have been concealed

1857. there, as an explosion took place, scorching the horse severely, which galloped off direct to the regimental sick-horse lines, where it died a day or two later.

Malleson thus writes of the work of the cavalry: "The Cavalry were ever ready, always on the alert, prompt to take advantage of every opening, and had been engaged in covering advanced movements, in keeping up long lines of communication, and in repelling the counter-attacks of the enemy on our right. Such services were necessary to the safety of the force, and though they might be less prominent, they were not inferior in value to those of the other arms. The names of Little, Ouvry, Probyn, Watson, Younghusband, and Gough, still live in the memory of those who served under them."

On the 27th, the regiment marched with the convoy which took the wounded, the ladies, children, and treasure to Cawnpore: arriving there on the 30th.

On the 6th December, Sir Colin Campbell, having got rid of the convoy, decided to attack the rebels outside Cawnpore, on the road to Calpee. The cavalry were under Colonel Little, and Sir Hope Grant was second in command to Sir Colin Campbell. The enemy were engaged, defeated, and put to flight; being for two miles pursued by Bourchier's battery, Hope Grant, and his Staff alone. The cavalry and horse artillery had been taken a wrong route by a guide, and did not come up for some little time. When they did, the order for further pursuit was at once given, and the cavalry spread quickly over the plain in skirmishing order, headed by Sir Hope. During this pursuit the following quaint incident has been written to me by Private D. Wareham, 9th Lancers (now a pensioner in Chelsea Hospital), which I give in his own words: "In the midst of the fight Sir C. Campbell sent an A.D.C. to Sir Hope Grant to bring up the 9th Lancers, and away we went at a dashing pace down the Calpee Road. It was a very dusty day, and there was a well on the left-hand side of the road covered over with long grass, looking to horse and rider as perfectly sound ground. When I came up to it, without any warning down I went, like a flash of lightning; it was a big well at the bottom, and very deep, about thirty-six feet, and that is what saved me; for Captain T. W. Hutchinson, who was riding about twenty yards behind me, while I and my horse were underneath the water, also fell down the well, and there we were, two men and two horses in this well. You can imagine the state of mind we were in until we were both pulled up by the Naval Brigade." 1857.

The enemy were pursued up to the fourteenth milestone on the Calpee Road; many guns were captured, and the force did not get back to their camp which was the ground where the battle had been fought early in the day, until midnight. Here they bivouacked; the night was cold, and the troops had little to eat. The Commander-in-Chief (Sir Colin) was the most thorough soldier of the lot; when his troops had to sleep in the open, he always made a point of stopping with his men.

The left wing and centre of the enemy's army made good their retreat towards Bithoor through an inoperative action under Mansfield, Chief of the Staff.

On the 8th, the regiment with a mixed force under Sir Hope Grant, moved off at 1.30 P.M. in pursuit of the Gwalior rebels. By careful inquiry on

1857. the road he found that they had gone towards the Seraighat Ferry, about three miles from Sheorajpur. He halted his troops, had them fed, and at nightfall set off again. He came up with the enemy preparing to cross the Ganges; after a short but sharp engagement, chiefly artillery; the rebels were absolutely routed, and fifteen guns were captured. Sir Hope states: "My gallant regiment, the 9th Lancers, was in support of the guns." Sir Hope was struck with a grape-shot on the foot, but luckily without being wounded.

The following extract from 'Recollections of a Winter Campaign in India, 1857-58,' by Captain Oliver J. Jones, R.N., may be of interest. "The Grand Trunk Road (three miles from Bithoor). It was here where I was first introduced to Sir Hope Grant, whose kindness and friendship to me, whenever afterwards I had the good fortune to meet him, I have so much reason to rejoice at and be thankful for. The 9th Lancers, of which he was colonel, was a perfect pattern of efficiency and order.

"After the long campaign they had been through, and the immense quantity of good service they had done, they used to turn out as well dressed and clean as if they were going to be inspected at Wormwood Scrubs. Every man was dressed in his uniform, and there were no yellow-leather boots pulled up over their trousers and all sorts of motley costume, which obtain in many regiments, and which meet with so much approbation in the present day, but which tend very much to lower the *esprit de corps* and *morale* of regiments. One thing certain, that with the proper strictness about uniform kept up in the 9th Lancers, there was no happier corps, no corps in which discipline was maintained with less punishment; and one only had to see them in the field to be satisfied that few regiments could equal, none could surpass, the gallant, the dashing, always-to-the-front 9th Lancers, every man of which, from the colonel to the trumpeter, looked like a gentleman." 1857

On the 11th, Sir Hope Grant was directed to at once march on Bithoor, and destroy the town and residence of Nana Sahib. Finding the place evacuated, he blew up the temple and burnt the palace. Many quaint articles of pure gold and silver were found in a well near the palace. Owing to want of carriage, which had been largely used for transport of wounded, ladies, and children, the army was inactive, but the regiment appears to have returned to Cawnpore on the 14th.

On the 16th, a one-innings cricket-match was played between the regiment and the artillery at Nawab Gunge, just outside Cawnpore on the road to Bithoor, the scores being—9th Lancers, 190 runs; R.A., 140. On the 19th, the Headquarter squadron, with Anson, Fawcett, Bell-Martin, and Sitwell (just joined), was at Nawab Gunge.

The other three squadrons, under Steele, Coles, and Johnson, left this day with, or to join, Walpole's force marching on Etewah, which place was blown up. This force joined Seaton's on the 3rd January 1858 at Bewar, on the road to Futtigarh.

About the 15th or 16th, a detachment of the regiment, with a squadron of the Carabiniers (140 in all), was employed with Seaton's column, and was engaged in the action of Gungaree, or, as it is sometimes quoted,

1857. that of Khasganj. The enemy was supposed to be thirteen miles distant. Hodson, and Light of the Bengal Artillery, who had gone forward to reconnoitre, soon galloped back and gave the alarm. In a moment all was bustle, the infantry turned out in their shirt-sleeves, the cavalry were in the saddle in three minutes, the gunners as quick.

The Carabiniers and Lancers were on the extreme right, Hodson's Horse on the left. Our line being threatened by the enemy's guns, Wardlaw of the Carabiniers, with his squadron, charged down on them and captured them. Out of the few Carabinier officers engaged, three—Wardlaw, Hudson, and Vyse—were killed. Lieut. and Brevet-Captain Head of the regiment, leading his handful of men, was dangerously wounded. The Cavalry were then led by Lieut. Russell, the only officer left, and cleared out the enemy.

On the 17th, these detachments were again engaged under Seaton near a village called Pattiali, when, after the defeat of the enemy, they, with Hodson's Horse and the Royal Artillery, pursued the rebels for seven miles, and took thirteen guns, also the elephant, the silver howdah, and the sword of Hakhim, Commander-in-Chief of the Nawab of Farrakhabad.

A curious accident happened to Captain Frank Grant about this date. He was travelling in a gharri, when the tyre of one of the wheels came off as it was going down a hill. He put his head out to see what was the matter, and the loose tyre caught him right across the face, cutting his nose and inflicting a severe gash on his forehead.

On the 2nd January, the Headquarter squadron was engaged, together with Probyn's and Younghusband's Sikhs, at the action of Goorsaigunj Bridge, when the Naval Brigade guns, under Vaughan and Peel, made such excellent practice. The cavalry under Hope Grant, caught the enemy just as they were emerging from the village, and pursued them for several miles. Their rout was complete; eight guns and other things were captured, amongst them the Queen's colour of the 10th Oude Irregular Infantry, which was presented to Sir Colin Campbell. Unfortunately, Younghusband here got his death-wound. Sir Hope was hit in the side by a spent shot, but except for a severe shock, was no worse. 1858.

Sir Hope Grant about this period relates one or two incidents which may be of interest.

His orderly, Corporal Caine of the regiment (a fine rider, and a plucky young fellow), always carried a bamboo lance for Sir Hope's use. One day he saw a rebel a little ahead of Sir Hope with his musket cocked and levelled, defying any one to approach. Caine brought his lance to the "engage" and rode at him. When about five yards off the rebel fired; the musket missed fire, and Caine ran him through the body.

"Coming back to camp one day after dark, to our surprise we found a gun we had captured earlier in the day going quietly along drawn by bullocks in charge of two rebels, one on the limber, with tulwar and musket, acting as guard. Lieut. Bell-Martin rode at him, and when within a few yards, they exchanged shots, Martin with his revolver, the rebel with his musket. Both missed. The rebels were killed, but the gun tumbled

1858 into a ditch with the bullocks on top. Peel of the Naval Brigade was by, and extricated it."

On the 3rd, the regiment was again all together. On the 27th, two squadrons of the regiment (Steele's and Johnson's) were engaged at the action of Shamsabad, and with Hodson's cavalry, pursued the enemy eight or nine miles, capturing four guns. Johnson was commanding the cavalry, Steele the two squadrons, Goldie and Willis in command each of a squadron. Goldie's squadron charged into 1500 of the enemy, and killed 300. Steele here had a narrow escape. He was attacked with a tulwar by a rebel on foot, and his sword-arm and reins were cut. The man aimed a blow at his head, but his horse spinning round, it fell on his thigh, which was ripped up. The men were ten hours in the saddle before the attack, and nineteen before the pursuit was abandoned. Lieut. Goldie (whose father and two sisters had been murdered at Futtigarh when the Mutiny broke out) was recommended for the V.C. Regimental-Sergeant-Major David Spence was awarded the V.C. for having rescued from the rebels Private Kidd, who was severely wounded.

On the 4th February the regiment arrived at Cawnpore from Futtigarh.

On the 8th, Sir Hope, in the absence of Sir C. Campbell, was placed in command of the whole force between Cawnpore and Bunnee. He was also appointed a Major-General in the army for distinguished service, and General of the cavalry division, Colonels Campbell and Little of the regiment being nominated his brigadiers. Their brigades on the 11th February were composed as follows:—

Campbell's:—7th Hussars, 2nd Dragoon Guards, Volunteer Cavalry, Hodson's Horse, detachments of 1st and 12th Irregular Cavalry, with other small lots, equal to about one regiment. 1858

Little's:—9th Lancers, Probyn's Horse, Watson's Horse, Bruce's Horse, Wales's Horse.

68 siege-guns, 50 field-pieces.

The following is a copy of the Order promoting Sir Hope Grant a Major-General:—

General Order.

HORSE GUARDS, *Feb.* 20, 1858.

The Queen has been graciously pleased to command that Colonel Sir James Hope Grant, K.C.B., of the 9th Lancers, be promoted to the rank of Major-General in the Army, in consequence of his eminent services in command of the Cavalry Division at the Siege of Delhi, and in that of a Divison at the Relief of Lucknow, under General Sir C. Campbell, &c.

About this time he had a distinguished service pension of £200 a year granted to him, but by his promotion to major-general's rank, he lost his purchase-money.

On the 23rd February, the force under Sir Hope, was engaged in the capture of the fort of Meanggunge. The fort was taken by the 53rd Regiment, headed by Captains Jones, R.N., and Anson (Hope's A.D.C.). On the flight of the rebels, Sir Hope says: "The 9th Lancers ran them through: the 7th Hussars and Irregulars cut them down." The squadrons of Fawcett and Coles did good work, the latter having his horse shot, and Lieut. Evans was particularly active among the rebels with a hog-spear. In a private letter of Sir Hope's, dated Mohan, 25th February 1858, he says:

1858. "I have recommended an excellent young officer of the name of Roberts for the V.C." (the late Commander-in-Chief).

On the 2nd March, the regiment with the army under Sir C. Campbell, arrived at the Dilkoosha, Lucknow; having had a skirmish on the way which nearly proved expensive in the way of commanders. Sir Colin Campbell, Sir Hope, and Colonel Little had gone forward to see to the removal of two apparently deserted ammunition-waggons. All of a sudden a heavy fire was opened upon them, and having only a small escort, the trio withdrew; but Colonel Little was so badly shot in the elbow that he had to relinquish his brigade command, which was taken up by Colonel Charles Hagart of the 7th Hussars. The regiment did not get into camp until 7.30 P.M., having marched at 4 A.M.; fifteen hours in the saddle.

On the 6th, a squadron of the regiment, with two squadrons of the "Bays," was engaged chasing the rebels after the crossing of the Gomtee, below the Dilkoosha. Private William Goat of the regiment was awarded the V.C., for having endeavoured to rescue from the enemy, the body of Major Percy Smith of the "Bays," who was killed in this action.

Sir Hope Grant thus describes his first meeting with Dr. Russell (war correspondent to the *Times*) at a durbar: "My fighting turn-out was however very indifferent, so I got behind the crowd in the durbar tent, and conversed with a stoutish gentleman in Hessian boots and frock-coat, who was a stranger to me. . . . His remarks were very clever, and much to the point."

On the 19th, two squadrons were engaged under Sir Hope with Sir James Outram's division, in dislodging the rebels from Moosa Bagh, where they had congregated after the fall of Lucknow. The squadrons followed up the rebels for four miles and took six guns. Captain T. Hutchinson (of the well incident on 6th December 1857) received his death-wound in this engagement. He was in the act of cutting down a rebel, who, getting beneath his sword-cut, pierced him in the eye with a poisoned spear, the thrust penetrating to the brain. He died two or three days after. Sir J. Outram states: "The conduct of the officers and men of the 9th was most gallant, as they undauntedly charged masses of the enemy." Troop-Sergeant-Major Rush and Private Robert Newell so distinguished themselves by attacking and cutting up a party of the enemy's infantry that they were both awarded the V.C. (*Note.*—This probably was in a combat with about fifty of the rebels, when Slade, Wilkins, and Banks of the 7th Hussars were cut down, and when Colonel Hagart saved Banks's life for the time being. He died of his wounds later, having lost a leg and an arm.) Coles's squadron captured twelve guns. 1858.

On the 7th April, the regiment left Lucknow with Brigadier-General Walpole's division, proceeding into Rohilcund, and was present at the attack on Roorja (Ruiya) Fort; as also on the 22nd, at the artillery and cavalry action at Sirsa, near Alleegung, when Captain Alfred Wilkinson with his troop captured three guns.

The regiment moved with the force under Sir Colin Campbell for the capture of Bareilly, and with the 2nd Punjab Cavalry, detachments from the Lahore

1858. Light Horse, 1st and 5th Punjab Cavalry, and 15th Irregulars, under Colonel Hagart, 7th Hussars, formed the 2nd Cavalry Brigade. The town was besieged on the 5th May. Owing to the great heat, after a partial capture of the place the army halted for the day, and on the 6th, Sir Colin found the place had been evacuated.

A portion of the regiment was engaged at Panhat on the 18th, under the Commander-in-Chief, and on the 24th pursued the rebels as far as Mahoom Dee, a distance of twenty miles, expelling the rebels from Rohilcund.

On the 7th June, the regiment marched for Umballa, arriving there on the 24th, having been thirteen months under canvas. A few days after its return there, it was inspected by Major-General Sir Robert Garrett, K.C.B., who said "he had come prepared to make many allowances, which he found were not required, as the appearance of the regiment was such as would at any time have been highly creditable."

In the early part of 1858, Sergeant Hartigan of the regiment, who had earned the V.C. for gallant conduct when the main-guard was attacked near Agra on the 10th October 1857, was employed helping Captain R. J. Meade to drill and instruct his newly formed regiment, which afterwards did such good work. Sergeant Hartigan was eventually given a commission in the 16th Lancers.

Early in September the regiment was ordered to march towards Cawnpore for "service" again. Before leaving Umballa a ninth troop was added, thus making, with the depôt troop, a total of ten.

On the 5th October the regiment reached Delhi, where their old friend the ex-King thereof with his family, was handed over to their charge to escort to Allahabad *en route* to Peru, to which place he had been banished. From Allahabad two squadrons escorted him as far as Mirzapore, from whence he proceeded to Calcutta. 1858.

The regiment reached Fyzabad on the 24th November; on the 25th it was present when the passage of the river Gogra was forced and the enemy utterly routed. After this the regiment underwent severe work, with long and rapid marches, occasionally having a brush with the enemy, the final one being at Kumbda Kothi on the 4th January 1859. 1859.

This completed the services of the "Ninth" in the suppression of the Indian Mutiny, the total loss during the period it was engaged—8th June 1857, to January 1859—being 5 officers and 143 men. It marched 3000 miles, and was the only regiment present at the three main achievements of the war—the Siege of Delhi, the Relief, and the Siege of Lucknow.

The regiment being under orders for England, was re-formed into eight troops, and handed over the horses and 147 volunteers, to other corps at Cawnpore. It marched *viâ* Allahabad to Calcutta, and embarked for England on the 1st May.

On its departure the following farewell and highly complimentary Order was issued by the Commander-in-Chief in India:—

The 9th Lancers began their fine career in India with the Gwalior Campaign, including the battle of Punniar; after which they participated in the Sutlej and Punjab Campaigns with the

1859. battles of Sabraon, Chillianwallah, and Goojerat. In 1857 and 1858 they were most prominent at the Siege of Delhi, having served and driven guns, in addition to their other duties during that trying time.

At the Relief of Lucknow, the Battle of Cawnpore, the Siege of Lucknow, the Campaigns in Rohilcund and Oude, the Regiment was most prominent.

Such is a very slender sketch of the services performed by the above Corps.

The following Gazette-Extraordinary was also published by the Viceroy and Governor-General :—

Number 502.

FORT WILLIAM, 13*th April* 1859.

After a long and brilliant career in India, the 9th (Queen's Royal) Lancers are on the eve of leaving Calcutta for England. From the beginning of the hostilities in which the British Army in India has recently been engaged, this Regiment has held a foremost place in the contest. Its name is inseparably connected with the Siege of Delhi, during which the services rendered by officers and men were various and most important, and in the latter campaigns of Rohilcund and Oude it has closed its term with much honour. In taking leave of the 9th Lancers, the Viceroy and Governor-General in Council desires to offer to this distinguished Regiment his hearty thanks, not only for its good services before the enemy, but for the admirable example which it has presented to the young soldiers of the Indian Army, by its perfection of discipline, conduct, and efficiency. The Regiment will be saluted by the guns of Fort William on its departure.

By Order of His Excellency the Viceroy and Governor-General of India in Council.

(Signed) R. H. J. BIRCH,
Major-General,
Secretary to the Governor of India.

The regiment embarked at Calcutta 403 all ranks, and arrived off Spithead on the 4th September, having

been 17 years and 122 days in India. The head- 1859.
quarters proceeded to Exeter, the inhabitants of which place, on the 13th October, invited the whole of the regiment to an entertainment in honour of their services during the Mutiny.

CHAPTER VII

1860–1885

1860. On the 4th January, Regimental-Sergeant-Major David Spence, Troop-Sergeant-Major David Rush, Lance-Corporal J. Freeman, and Privates W. Goat and P. Donohoe, proceeded to Windsor; and were there presented by Her Majesty the Queen, with the Victoria Cross, which they had won in the Mutiny.

The following is a complete list of the officers, non-commissioned officers, and men awarded this decoration :—

Lieut. A. S. Jones	Budlee-ki-Serai	8th June 1857.
Lieut. R. Blair (attached)	Boolundshuhur	28th Sept. 1857.
Private P. Donohoe	"	" "
Lance-Corporal Freeman	Agra	10th Oct. 1857.
Private W. Goat	Lucknow	6th March 1858.
Private Thos. Hancock	Delhi	19th June 1857.
Sergeant H. Hartigan	Agra	10th Oct. 1857.
Private Henry Jordan	Boolundshuhur	28th Sept. 1857.
Trumpeter R. Kells	"	" "
Private R. Newell	Lucknow	19th March 1858.
Private J. Purcell	Delhi	19th June 1857.
Private J. R. Roberts	Boolundshuhur	28th Sept. 1857.
Troop-Sergeant-Major D. Rush	Lucknow	19th March 1858.
Regimental-Sergeant-Major D. Spence	Shamsabad	27th Jany. 1858.

On 24th March, it was ordered that gauntlets be worn on mounted parades only.

On the 12th June, a revised edition of the Standing Orders was issued by Colonel A. Little, C.B. 1860.

In September the regiment moved to Aldershot.

The Annual Regimental Dinner was established this year, but there is, however, no record of officers dining at it until 1866. (See Appendix IX.)

The regiment moved to Brighton and Shorncliffe in August—six troops to the former place, two to the latter. 1862.

By General Order 834, dated September, the regiment was authorised to wear on its colours and appointments the words "Delhi" and "Lucknow." 1863.

In March, the regiment moved to Newbridge, and in September to Dublin. 1864.

On the 5th February the Colonel of the regiment, General Sir James Wallace Sleigh, K.C.B., died, and was succeeded by Lieut.-General Sir J. Hope Grant, K.C.B. 1865.

In April it moved headquarters and six troops to Dundalk, a troop each to Belfast and Belturbet.

From July to October, the regiment was at the Curragh, moving in the latter month to Dublin. 1866.

In October, the headquarters moved to Cahir, with out-quarters at Cork, Waterford, Clonmell, &c. 1867.

In October the regiment moved to Newbridge. 1868.

1869. On the 19th May, Lieut.-Colonel C. P. Johnson retired from the army, and was succeeded in the command of the regiment by Lieut.-Colonel the Honourable Ivo-de-Vesci W. Fiennes.

In July, it returned to England, and was quartered at Hounslow, out-quarters at Hampton Court and Kensington.

1870. The regiment moved to Aldershot.

1871. The following was the establishment fixed for the various ranks in February of this year: 1 colonel, 1 lieutenant-colonel, 1 major, 8 captains, 8 lieutenants, 4 cornets, 1 paymaster, 1 adjutant, 1 riding-master, 1 quartermaster, 1 veterinary surgeon, 1 regimental-sergeant-major, 1 bandmaster, 1 quartermaster-sergeant, 1 trumpet-major, 1 paymaster-sergeant, 1 armourer-sergeant, 1 saddler-sergeant, 1 farrier-major, 1 hospital sergeant, 1 orderly-room clerk, 8 troop-sergeant-majors, 1 sergeant-instructor of fencing, 1 sergeant-cook, 24 sergeants, 8 farriers, 32 corporals, 8 shoeing-smiths, 2 saddlers, 1 saddle-tree maker, 504 privates.

1872. In May, the regiment moved to Woolwich, and in September marched to York.

1873. In August, it took part in the manœuvres held at Cannock Chase.

In November, Colonel Fiennes retired on half-pay, and was succeeded in command of the regiment by Lieut.-Colonel H. Marshall.

In June, it was decided to retain the blue frock-coat as an article of "undress" uniform in place of the patrol-jacket. 1874.

In July, the regiment marched from York to Colchester.

In January, it was placed under orders for India, and the horses were handed over to the depôt of the 5th Lancers. A depôt troop was left at Canterbury, and on the 8th, the regiment embarked on H.M. troopship *Euphrates*, strength as follows: Field officers, 2; captains, 5; subalterns, 9; staff, 5; staff-sergeants, 10; sergeants, 24; farriers, 7; trumpeters, 5; corporals, 22; privates, 366. The troopship sailed on the 9th, and reached Bombay on the 10th February. 1875.

On the 16th, the regiment left Deolali for Mean Mir which was reached on the 24th, and proceeding by march route arrived on the 9th March at Sialkote, where the horses of the 5th Lancers were taken over.

In December, a regimental refreshment-room was established, with the following object: "To supply to the men of the regiment the best articles that can be procured, at low prices; and to provide a place where every man can, at any time in the day, procure a good and comfortable meal."

On the 17th January, a detachment of the regiment consisting of 100 men, under Captain Bushman, the Honourable H. Legge, Lieuts. J. A. Stewart-Mackenzie and Butson, with Riding-Master J. Marshall and the band, proceeded to Jummoo; to act as escort to Field-Marshal H.R.H. the Prince of Wales. 1876.

1876. In December, the regiment sustained a great loss by the death of Captain W. C. Clayton, who was killed at Delhi on the 25th of the month while playing polo. Lieut.-Colonel Marshall, in the Regimental Order announcing his death, said: "The zeal and energy of the late Captain W. C. Clayton for the good of every one, and the general welfare and interest of the regiment to which he was so proud to belong, has secured him a place in the affection of every officer, non-commissioned officer and private in the 9th Lancers." Captain Clayton was buried in the old cemetery below the Ridge at Delhi—which was specially opened for the purpose—alongside of another 9th officer, Brevet-Lieut.-Colonel R. Yule, who was killed in 1857.

1878. On the 19th October, the regiment marched from Sialkote under Major R. Cleland, for service in the field, and reached Nowshera, where a halt was made. Martini-Henry carbines were here issued to the regiment, and the old pistols were sent to the arsenal at Ferozepore. It was put through a brief course of instruction with its new weapon by the adjutant, Lieut. E. B. M'Innis, and finally, after a short stay at Tarvoo, joined Sir Frederick Maude's Division at Jumrood, at the mouth of the Kyber Pass. The division proceeded by way of Ali Musjid and Dakka, to Narkoo, on the right bank of the Kabul river, where it remained until the beginning of June 1879.

1879. On the 14th January, the first squadron under Captain Apperly, with Captains J. A. Stewart-Mackenzie and Butson, left Tarvoo to join the Kurrum Valley Field Force under Sir Frederick Roberts, relieving a squadron of the 10th Hussars. 1879.

On the 2nd June, the treaty of Gundamuk was signed, and hostilities being over, the regiment less the first squadron which remained with the force above mentioned returned to Sialkote, which was reached on the 22nd of the same month.

On the 23rd August, Captain, Lord William Beresford was awarded the V.C. for gallantry during the Zulu war.

On the 3rd September, the embassy to Kabul under Sir Louis Cavagnari was massacred. Hostilities being again resumed on the 17th, the headquarters and two squadrons under Captain Stewart-Mackenzie left Sialkote for Kabul, which was reached on the 4th November. Lieut.-Colonel Cleland, who had succeeded Lieut.-Colonel Marshall in the command of the regiment, was away on leave, but joined in a few days.

To return to the first squadron, which had in January been detached from the regiment. On the 22nd March, it entered the Kurrum Valley and camped near Kurrum Fort, from whence it moved on 4th May, and with one or two halts reached Ali Kheyl on 23rd.

On the 24th May, the Kurrum Valley division, numbering about 9000, mustered for Queen's Birthday parade. The following troops formed the division: First squadron 9th Lancers, 14th Bengal Lancers, 12th Bengal Cavalry, 5th Punjab Cavalry; F Battery, A Brigade, R.H.A., Royal Artillery C 4 and G 3, and Nos. 1 and 2 Mountain Batteries; the 92nd and 72nd

1879. Highlanders, the 67th Regiment, 23rd Pioneers, 5th Goorkhas, 11th, 21st, and 28th Regiments Native Infantry, and the Bengal Sappers and Miners.

During the months of July and August, cholera was bad in the artillery ranks and somewhat bad in the squadron.

On the 10th September, the squadron returned to Kurrum Fort, and on the 23rd the order was given for an advance on Kabul. On the 27th the squadron formed the advance-guard; two squadrons of the 5th Punjab Cavalry and the 28th Native Infantry the rear-guard. While crossing Shatar-gardan it was engaged with the Monguts at Taga Thana, but supported by the rear-guard and about thirty of the 92nd Highlanders, who had come out as a guard of honour to meet General Roberts, the enemy was driven off.

On the 6th October, a party of twenty men under Captain Apperly moved from Charasia at daybreak to reconnoitre the pass. They were almost immediately fired upon, and commenced the action of Charasia, which ended at sunset by our troops being in possession of all the main positions in the vicinity. In the main action, two squadrons composed of detachments from the 9th Lancers, 12th Bengal Cavalry, and 5th Punjab Cavalry, were commanded by Major Hammond of the latter regiment, but had not much to do.

On the 8th, a cavalry force composed as follows—102 of the 9th Lancers, 140 of the 5th Punjab Cavalry, 260 of the 12th Bengal Cavalry, and 220 of the 14th Bengal Lancers, under General Massy—was sent to cut off the retreat of the enemy, who was said to have abandoned Sherpur. Sherpur was found deserted, and

seventy-five guns were taken, which had been left 1879.
behind by the enemy, who was seen on the Asmai Height overlooking the old Afghan quarter of Kabul. The object of the cavalry now was, to pass into the Chardeh Valley and head off the enemy from the road leading to Ghazni, &c. An infantry and artillery action was carried on from 2 to 6 P.M., but no decisive result was this day obtained. The troops bivouacked where they stood, and had no rations.

On the 9th, the squadron and native cavalry started off in pursuit of the enemy who had vacated Kabul during the night, leaving their guns, tents, camp equipage, &c. The force left camp at 4.30 A.M., and did not return till 9 P.M., having gone more than twenty-one miles out in pursuit of the enemy, who however could not be caught, with the exception of a small party of about twenty, who were cut up by the 5th Punjab Cavalry. Kabul was now ours, and on the 13th, the whole of General Roberts' force marched through the city.

On the 27th, twenty-three young horses, which formerly belonged to the Ameer's army, were sent to the first squadron.

On the 4th November, the headquarters, and the other two squadrons of the regiment under command of Lieut.-Colonel Cleland, joined the first squadron at Sherpur cantonments.

On the 22nd, the second squadron under Captain J. A. Stewart-Mackenzie, was engaged in the action on the Bamian Road and the destruction of Bahadar Khan's Fort.

On the 1st December, a troop of the regiment, with

1879. two guns, G 3 Royal Artillery, and a troop of the 5th Punjab Cavalry, under the command of Major Hammond, of that regiment, escorted Yakub Khan about twelve miles on his way to India, as an exile.

On the 7th, a troop of the regiment formed part of the escort which guarded the late Commander-in-Chief of the Ameer's army, a prisoner to India.

On the 9th, a squadron of the regiment (F and D troops) accompanied a small force under Brigadier-General Macpherson, and encamped that night at Aoshaar, taking part next day in the defeat of a force under Mir Butcha, near Kila Karez.

On the 11th, acting as escort to four horse-artillery guns under Major Smith Wyndham, this squadron moved off early in the morning, and later was overtaken by Brigadier-General Massy with another squadron of the regiment under Lieut.-Colonel Cleland, and forty-four of the 14th Bengal Lancers under Captain Phillip Neville. It was intended by Sir F. Roberts that this force of artillery and cavalry should act in co-operation with General Macpherson's infantry.

General Massy moved direct across country to the Ghazni Road; and when near Kila Kazi, a large body of the enemy (eventually estimated at 10,000, under Mahomed Jan) was reported by the advance-guard under Captain Bloomfield Gough.[1] This force bore down on General Massy, and was too strong for his small artillery and cavalry force, unsupported by Macpherson's infantry; with which a junction had not

[1] This gallant officer met with a sudden death in June 1904, being thrown out of his pony-cart and killed on the spot.

been made. The guns had to retire, until the enemy 1879.
was within carbine range of them. At this point Sir F. Roberts appeared on the scene, and it being most important that Mahomed Jan's advance on Kabul should be delayed, he gave the order to the cavalry to charge; and 126 of the regiment and 44 of the 14th Bengal Lancers, led by Lieut.-Colonel Cleland, charged into the mass of Afghans, being joined by Captain Gough's troop, who, as above stated, was acting as advance-guard on the right of our line. They were received with a terrific fire, and Sir F. Roberts ('Forty-one Years in India') thus describes the incident: "The charge was led by Lieut.-Colonel Cleland and Captain Neville, the former of whom fell dangerously wounded; but the ground, terraced for irrigation purposes and intersected by nullahs, so impeded our cavalry that the charge, heroic as it was, made little or no impression upon the overwhelming numbers of the enemy. The effort, however, was worthy of the best traditions of our British and Indian cavalry, and that it failed in its object was no fault of our gallant soldiers."

The guns, in a further retirement, had got into difficulties, and with a faint hope of saving them Captain Stewart-Mackenzie, with the now rallied squadron, was ordered to make a second charge. This was done, but to little or no purpose.

Our loss was heavy. Lieuts. Ricardo and Hearsy, and 16 men and 34 horses, were killed; Lieut.-Colonel Cleland was severely wounded by shot and knife, and died of his wounds on 7th August 1880; thirteen men and 37 horses were wounded. Captain Mackenzie's

1879. horse was shot and fell on him, he being extricated with much difficulty.

The Rev. J. W. Adams, Chaplain to the Force, was awarded the V.C. for succouring a wounded man of the regiment. He dismounted and endeavoured to lift the man on to his own charger. The mare broke away and was never seen again; he however stood by his man until able to hand him over to some of his own comrades. He also extricated two more men of the regiment, who were pinned underneath their horses at the bottom of a ditch. He died on the 20th October 1903 at Oakham.

On the 13th, while engaged in attending the burial of officers and men who were killed on the 11th, an order was received for the regiment to turn out at once. With the 5th Punjab Cavalry, a squadron of the 14th Bengal Lancers, and the Guides Cavalry, they fell on the enemy, who had been driven off the Takht-i-Shah Peak on the Siah Sung Heights, and caught them on both flanks, scattering them all over the plain. One brilliant charge by the regiment, and the Guides, was most disastrous. Captain Butson (in command of the regiment) was shot through the heart; Captain Scott-Chisholme was shot through the thigh, the flash of the rifle burning his clothes, so close was the discharge of the rifle. Although most severely wounded, he remained in the saddle and brought the regiment out of action. Lieut. Trower was wounded, Sergeant-Major Spittle, four men and four horses were killed, and seven men were wounded.

The regiment now returned to the cantonments at Sherpur, and on the 23rd, assisted in repelling an attack of the enemy thereon, in which they were utterly defeated, being pressed back in the direction of Kabul. 1879.

On the 24th, the regiment, with the 14th Bengal Lancers, turned out to pick up stragglers from the fight of the previous day. The ground was covered with deep snow, and was so slippery that progress was difficult, and the men had to dismount and lead their horses.

On the 29th, a squadron of the regiment with other troops, went to the Chardeh Valley, where the engagement of the 11th December took place, in order to try and recover some of the bodies of men killed that day. The ground was so slippery from ice and snow that many horses fell and rolled down the hillside.

On the 2nd February, a great battle took place 1880. (with snowballs). The regiment, with the Artillery and the 72nd Highlanders, stormed a position on the Burmura Heights, which had been occupied by the 92nd and 67th, and succeeded in driving them from it.

On the 28th February, Major H. A. Bushman, having rejoined from special service in South Africa, assumed command of the regiment.

In April, a draft of 1 officer and 83 men, volunteers from the 17th Lancers, joined the depôt at Sialkote.

On the 16th, a squadron under Major Abadie moved with a force of 4000 strong, commanded by Major-General J. Ross, conveying supplies for Sir Donald Stewart's column.

On the 25th, a squadron of the regiment was employed during the battle of Charasia, near which place the Logaris had surrounded a force under Colonel

1880. Jenkins. The cavalry brigade and four horse-artillery guns, under General Hugh Gough, supported a force under General Macpherson, which was sent to his aid. The enemy was broken and scattered, and was pursued by the cavalry and horse artillery for four miles.

On the 8th August, Major and Brevet-Lieut.-Colonel H. A. Bushman was promoted to the command of the regiment vice Lieut.-Colonel Cleland.

On the same date the regiment left Kabul on the march from that place to the relief of Candahar, which was closely invested by the victorious army of Ayub Khan, who had defeated our troops under General Burrows on the 27th July 1880 at Maiwand, forty-five miles from Candahar.

The cavalry brigade under Brigadier-General Hugh Gough, V.C., was composed as follows :—

9th Lancers,	commanded by	Lieut.-Colonel Bushman.
3rd Bengal Cavalry	,,	Lieut.-Colonel Mackenzie.
3rd Punjab Cavalry	,,	Lieut.-Colonel Vivian.
Central India Horse	,,	Colonel Martin.

The whole force, 10,000 strong, was under the command of Sir Frederick Roberts.

The following appeared in the Field Force Orders of the 14th August :—

Lieut.-General Sir F. Roberts is sure that all ranks of the late Cabul Field Force will share the regret he feels at the death of Lieut.-Colonel Cleland of the 9th (Queen's Royal) Lancers. On the 11th December last, in the Chardeh Valley, this officer was dangerously wounded whilst gallantly leading his distinguished regiment against the enemy. From the effects of the wound Lieut.-Colonel Cleland died at Murree on the 7th inst., after many months of severe suffering. By the death of Lieut.-Colonel Cleland, Sir F. Roberts, in common with a large number of officers and soldiers, has lost a valued friend, whilst Her Majesty's army has been deprived of the services of a most promising and gallant officer. The Lieut.-General desires to express the deep sympathy he feels with the officers and men of the 9th Lancers in the personal loss they have sustained. 1880.

The cavalry headquarters reached Robat on the 27th August, and heliographic communication was established with the Candahar garrison.

On the 28th the main body reached Robat, and on the 31st the force camped under the walls of Candahar.

The days' marches from Sherpur to Candahar are of interest—

August.		Miles.
8	Sherpur to Charasia	11
9	„ Zahidabad	7
10	„ Zerghun Shahr	14
11	„ Padkhao Barak	18
12	„ Saidabad	25
13	„ Takia	12
14	„ Shashgao	17
15	„ Ghazni	13
16	„ Yergatta	20
17	„ Chardeh	12
18	„ Oba Karez	16
19	„ Mukur	15
20	„ Panjak	21
21	„ Garjui	18
22	„ Bala Kazai	17
23	„ Khelat-i-Ghibzai	17
24	Halted.	
25	„ Jaldak	16
26	„ Tirandez	16
27	„ Khel-i-Akhund	14
28	„ Robat	20
		319

1880.

August.		Miles.
29	Halted.	
30	Robat to Momand	11
31	„ Candahar	10

On the 1st September, the regiment took part in the battle of Candahar, when Ayub Khan's army was utterly routed, and his treasury and camp equipment, together with thirty-one guns and two of our 9-pounder horse-artillery guns fell into our hands. It, with the 3rd Bengal Cavalry, 3rd Punjab Cavalry, and Central India Horse, under General Hugh Gough, were to have moved direct to the west, crossed the Argandat river, and cut off the enemy's retreat westward, acting simultaneously with the infantry attack. Owing to the village of Gundigan having been occupied by the enemy during the previous night, General Gough had to take his brigade eight miles round before he could strike the river; and this detour caused the brigade to appear rather too late to make the pursuit of the flying Afghan army a rout. Once over the river, the brigade opened out across the Argandat valley and pursued the enemy for about ten miles, returning to camp outside Candahar, which was reached at 8.30 P.M., having been in the saddle twelve hours.

From the 2nd September to 20th November, the regiment was encamped near Candahar. On the 25th September it was attached to the Bombay division, the cavalry brigade of which was commanded by Brigadier-General Wilkinson.

On the 25th November the march back to India commenced, the regiment having been relieved by the

13th Hussars, the two meeting on the way up on the 30th November at Segi. 1880.

On the 8th December, Pir Chokie was reached, and from here the regiment proceeded by squadrons by rail to Umballa, which was reached on the 13th, 15th, and 16th December. The depôt had already moved there from Sialkote.

This completed the services of the regiment in the Afghan campaign, during which, one squadron was in the field from October 1878 to December 1880, and the remainder, with a break from June to September 1879, a like period. Its loss, including those killed and those who died of wounds or exposure, was 4 officers and 51 non-commissioned officers and men. It was present at the battles of "Charasia," "Kabul," and "Kandahar," besides taking part in many slight brushes with the enemy. In commemoration of these services Her Majesty the Queen was pleased to approve of the above words, and "Afghanistan, 1878-80," being borne on the colours and appointments of the regiment.

On the 18th December, the following were presented by Lieut.-General W. Hughes, commanding the Sirhind division, with the medal for distinguished conduct in the field:—Regimental-Sergeant-Major R. Young, Troop-Sergeant-Major H. England, Corporals H. Loughead and S. Harsent, Lance-Corporal A. Long, and Private F. Druce.

In March, Captains Apperly, Stewart-Mackenzie, 1881. Scott-Chisholme, and Gough were granted the rank of brevet-major for their distinguished gallantry.

1881. In May, Lieut.-Colonel H. A. Bushman was appointed a C.B.

1882. On the 24th April, Lieut.-Colonel Bushman, C.B., presented the officers, non-commissioned officers and men with their medals for the Afghan campaign. A special bronze star had been awarded to officers and men who took part in the Kabul to Candahar march. There were 293 recipients of the Afghan medal, and 255 of the bronze star.

1884. On the 15th December, Colonel H. A. Bushman, C.B., gave up command of the regiment, taking an appointment on the Headquarter Staff of the army in India, and was succeeded by Lieut.-Colonel H. R. Abadie.

1885. On the 7th March, the regiment under command of Major the Honourable H. Legge, left Umballa for Rawal Pindi, to take part in a durbar held there by His Excellency the Viceroy and Governor-General of India (Lord Dufferin), at which the Ameer of Afghanistan (Abdul Rahman) was present. It, together with the 2nd Battalion Seaforth (78th) Highlanders and L Battery, A Brigade, R.H.A., formed the Viceregal Escort.

On the 1st April, affairs on the Afghan and Russian frontier having assumed a rather serious aspect (Punjdeh affair), the regiment was warned for service. The cloud however blew over. Between the 6th and 15th, the regiment took part in several reviews of the troops, and performed various escort duties to the Viceroy. On the latter date it returned by rail to Umballa.

On the 14th October, the regiment under command of Lieut.-Colonel H. R. Abadie, left Umballa by rail for Bombay *en route* for England, and sailed on the 28th in H.M. troopship *Crocodile*, strength as under:—Officers, 15; warrant officers, non-commissioned officers, and men, 414. Of these numbers, 2 officers and 87 non-commissioned officers and men had come to India with the regiment in 1875, and served there continuously. Major Apperly was left at Umballa to hand over the horses (394), saddlery, &c., to the 2nd Dragoon Guards. 1885.

On the 23rd November, H.M. troopship *Crocodile* reached Portsmouth, and on the following morning the regiment proceeded by rail to Shorncliffe, where the depôt—4 officers, 106 non-commissioned officers and men—had moved to from Canterbury on the 18th November, under Major B. Gough.

The regiment was again formed into eight troops.

CHAPTER VIII

1886–1899

1886. On the 18th May, H.R.H. the Duke of Cambridge, Commander-in-Chief, inspected the regiment on its return from service in India.

In June, a revised edition of the Standing Orders was issued.

On the 26th July, the Colonel of the regiment, General Sir A. Little, K.C.B., inspected it, and expressed himself as very well pleased with its appearance.

On the 6th August, the regiment left Shorncliffe by march route for York, arriving there on the 21st and 23rd August.

On the 28th September, a two-year-old cream-coloured filly from the Royal Stud at Hampton Court, was presented to the regiment by Her Majesty the Queen, to carry the kettle-drums. This kind gift of Her Majesty was to perpetuate a previous gift by H.M. King William IV. of two cream-coloured colts for the same purpose in 1832.

On the 12th October, a new pattern of sword was taken into use.

1887. On the 9th May, the regiment, together with the Yorkshire Hussars (Yeomanry cavalry) and the other troops in the York garrison, was reviewed on the Knavesmire by H.R.H. the Duke of Cambridge, Commander-in-Chief. 1887.

On the 19th May, Lieut. Lord John Thynne died in the garrison hospital at York from "compression of the brain," the result of an accident on the 16th inst. His horse slipped on the tram-lines opposite to the barrack gate, and threw him on to his head against a paved gutter.

On the 20th June, the regiment took part in the Jubilee Service, held in the Minster, at York; to commemorate the fiftieth year of the reign of H.M. Queen Victoria, and afterwards in a review of the troops held on the Knavesmire, under Major-General Daniell, commanding the Northern District.

By Army Circulars of this date, the establishment of the regiment was fixed as under:—1 lieutenant-colonel, 3 majors, 6 captains, 8 lieutenants, 3 second lieutenants, 1 adjutant, 1 riding-master, 1 quarter-master, 2 warrant officers, 10 staff sergeants, 8 troop-sergeant-majors, 8 farrier sergeants, 23 sergeants, 8 trumpeters, 32 corporals, 8 shoeing-smiths, 3 saddlers, 1 saddle-tree maker, 366 privates.

On the 30th June, Colonel H. R. Abadie relinquished command of the regiment, being succeeded by Lieut.-Colonel the Honourable Heneage Legge.

On the 1st July, the use of two shoe-cases was ordered to be discontinued, one only to be carried, containing a fore and hind shoe with nails.

On the 11th July, it was ordered as an experimental measure, that the use of cruppers be discontinued, 5 per cent being retained for bad-shouldered horses.

1887. On the 1st November, a new pattern of pistol (Enfield, Mark II., breech-loading revolver, with safety catch) was issued for use.

On the 5th November, Captain H.R.H. Prince Albert Victor of Wales, K.G., 10th Hussars, who had been attached to the regiment under date 30th September 1887, joined for duty.

1888. On the 11th, 12th, 13th, and 14th April, the regiment marched from York to Manchester, and Seaforth Barracks near Liverpool (two troops), being relieved at York by the 10th Hussars, when H.R.H. Prince Albert Victor of Wales rejoined his own regiment.

On the 8th May, an escort of 1 troop-sergeant-major, 1 sergeant, and 23 rank and file, under Lieut F. W. Duff, proceeded to Blackburn to escort H.R.H. the Prince of Wales while visiting that town.

On the 23rd, Major B. Gough's troop (H) marched to Denbigh in aid of the civil power, returning to Manchester on the 28th June. The following resolution, passed at the Court of Quarter Sessions for the County of Denbigh, was forwarded to the officer commanding the regiment:—

Resolved unanimously (on the motion of the Chairman of Quarter Sessions, B. T. Griffiths-Boscawen, Esq., seconded by the Lord-Lieutenant of the County, William Cornwallis West, Esq., M.P.): "That the Justices of the County of Denbigh, in Quarter Sessions assembled, hereby record their high appreciation of the services of the officers and men of the detachment of the 9th Lancers rendered to the Magistrates and Police of the County in maintaining order, and preventing to any extent breaches of the peace, at the recent Tithe Distraints. 1888.

(Signed) L. G. ADAMS,
Clerk of the Peace.

Dated 5th July 1888.

On the 20th October, a troop of the regiment together with the band, under Lieut. the Honourable C. Willoughby, took part in a procession through the City of Manchester, which was visited this day by H.R.H. Prince Albert Victor.

On the 5th July, a squadron under Major J. A. Stewart-Mackenzie marched to Birmingham, for duty in connection with the visit of H.M. the Shah of Persia to that city. 1889.

On the 15th, 16th, and 17th July, the regiment took part in the reception of, and furnished escorts for, H.M. the Shah of Persia when visiting Manchester; being thanked by the mayor and citizens thereof for their services on this occasion.

On the 21st August, a squadron of the regiment under command of Colonel the Honourable H. Legge, proceeded to Wrexham for duty in connection with the visit of H.M. the Queen to that town.

On the 5th, 8th, and 11th August, the regiment moved from Manchester and Seaforth Barracks to the Curragh for manœuvres, proceeding on the 16th and 17th September to Dundalk (headquarters) and Belfast. 1890.

On the 8th December, Lieut. J. L. de Laski died at Dundalk of enteric fever.

1890. On the 15th December, Colonel the Honourable H. Legge relinquished command of the regiment, and was succeeded by Lieut.-Colonel J. A. Stewart-Mackenzie.

1891. By Army Order 101, 1890, an active-service frock was ordered to be issued annually, a tunic biennially; and on the 20th May it was ordered that cotton head-ropes be substituted for collar chains.

On the 10th June, General Sir Archibald Little, G.C.B., colonel of the regiment, died at his residence, Upton House, Tetbury, and was succeeded by Honorary Lieut.-General, W. Drysdale, C.B. (London Gazette, 3rd July 1891).

1892. In February an enamelled iron water-bottle was substituted for the wooden one hitherto used.

From the 1st March, squadron organisation was established in cavalry regiments, and those of the regiment were thus formed:—

"A"	squadron,	G	and	H	troops.
"B"	,,	A	,,	E	,,
"C"	,,	C	,,	D	,,
"D"	,,	B	,,	F	,,

In June, the regiment moved to the Curragh from Dundalk, and in September proceeded to Newbridge.

On the 1st July, the Martini-Metford was substituted for the Martini-Henry carbine.

On the 1st August, the Webley was substituted for the Enfield revolver.

1893. In May, a new sword (cavalry pattern, 90) was issued.

From the 4th to 18th August, the regiment took part in the Curragh field column operations. 1893.

On the 13th October, a new pattern black-japanned ammunition-pouch was issued.

On the 28th October, Captain and Quartermaster John Forsythe retired from the army; having served in the regiment for thirty-four years, with an unbroken record of faithful, unfailing, and devoted service.

On the 13th November, a new carbine, .303 Martini-Metford, was issued to the regiment.

30th December. Many of those who have partaken of the unbounded hospitality of the late "Bill Beresford," and thanks to his thought and kindness, have had a "good time," whether at Simla, Calcutta, or elsewhere; may like to be reminded that on the above date, prior to his departure from India, where he had been for eighteen years—six as A.D.C. to Lords Northbrook and Lytton, and twelve as Military Secretary to Lords Ripon, Dufferin, and Lansdowne—he was entertained to dinner at the Town Hall Calcutta, by about 180 of his friends (soldiers, sailors, and civilians); at which the chairman said, "Bill Beresford has raised the office [of Military Secretary] to a science, and himself from an official into an institution."

On the 30th May, a new pattern field-service cap was issued. 1894.

In June and July, the regiment moved from Newbridge to Aldershot.

From the 31st August to the 21st September, the regiment was engaged in the cavalry manœuvres at Churn Camp, Berkshire.

1894. On the 15th December, Lieut.-Colonel J. A. Stewart-Mackenzie was granted an extension of one year in command of the regiment.

1895. In April, the Lee-Metford Magazine carbine was substituted for the Martini-Metford.

On the 15th December, Colonel J. A. Stewart-Mackenzie relinquished command of the regiment, being succeeded by Lieut.-Colonel B. Gough.

1896. On the 26th August, the regiment having marched from Aldershot, embarked at Woolwich on the Peninsular and Oriental steamships *Malta* and *Simla*, for service in South Africa, and sailed on the following day: 20 officers, 2 warrant officers, 41 sergeants, 34 corporals, 7 trumpeters, 466 privates, 413 horses; leaving a depôt of 2 officers and 106 rank and file at Canterbury.

On the 23rd September, it arrived at Durban, and proceeded at once by rail to Pietermaritzburg.

On the 11th December, two squadrons of the regiment under Major M. O. Little, proceeded from Fort Napier, Pietermaritzburg, to the Agricultural Show Grounds, and were there quartered until 26th April 1897; this being the period of "horse sickness" in Africa.

1897. In May, by Army Order 41, 1897, the four squadrons were reorganised into three Service Squadrons and a Reserve Troop. The "B" squadron was broken up to complete the increased establishment of the three squadrons, and to form the reserve troop.

The strength of each squadron and the reserve troop was as under :— 1897.

	Major.	Captain.	Subal-terns.	Sergeants.	Trump-eters.	Corporals.	Privates.	Horses.
Squadron	1	1	3	10	2	7	107	105
Reserve Troop		1	1	4	1	10	69	45

On the 28th May, the service squadrons marched for Ladysmith camp, where they arrived on the 2nd June. The reserve troop remaining at Pietermaritzburg.

On the 4th November, 2nd Lieut. W. S. Thorpe died at Rosboom, Natal; from a fracture of the skull, the result of an accident when bicycling.

On the 11th March, the regiment moved by rail from Ladysmith to Durban *en route* for India, handing over the horses to the 5th (Royal Irish) Lancers at the former place. It sailed on the following day in R.I.M.S. *Clive*, and reached Bombay on the 1st April. It disembarked on the 2nd, and proceeded by rail to Deolali *en route* for Muttra, which was reached on the 10th. The depôt, which had left Canterbury in September 1897 when the cavalry depôt there was abolished, was already at Muttra. The horses of the 5th Lancers were taken over. 1898.

In consequence of coming upon the Indian establishment, the regiment was formed into four squadrons, and in October the strength was fixed as follows:—29 officers, 2 warrant officers, 593 non-commissioned officers and men, and 525 horses.

From the 27th December, the regiment took part in the cavalry Camp of Exercise between Aligarh and Delhi, returning to Muttra on the 22nd January 1899. 1899.

CHAPTER IX

1899–1903

1899. September. In consequence of the strained negotiations between Her Majesty's Government and President Kruger over affairs general in South Africa, the regiment was ordered to mobilize and prepare for active service in that country.

On the 11th September, definite orders for service there were received.

On the 20th, the regiment left Muttra for Bombay, 16 officers, 475 rank and file, 518 horses; leaving a depôt at the former place, Captain H. Stanley in command.

The following officers embarked with the regiment:—

Major M. O. Little in command (Lieut.-Colonel B. Gough being in England on sick leave).
Major S. W. Follett, second in command.
Captain F. Lund.
Captain E. R. Gordon.
Captain H. E. Fiennes.
Lieut. F. Allhusen.
Lieut. H. Durand.
Lieut. E. H. Abadie.
Lieut. the Hon. D. Carleton.
2nd Lieut. S. Theobald.
2nd Lieut. J. S. Duckett.
2nd Lieut. F. W. Cavendish.
2nd Lieut. L. Sadleir-Jackson.
2nd Lieut. Lord F. Blackwood.
Quartermaster D. Laing.
Captain R. Forrest, Royal Army Medical Department.

Bombay was reached on the 24th, and by 1 P.M.

the "C" squadron on board the transport *Wardha*, and the "D" squadron with regimental staff on the *Nowshera*, were in the Bay *en route* for Durban. 1899. September.

The "A" squadron on board the *Nairung*, left Bombay the following day.

The *Wardha* reached Durban on the 9th, the *Nowshera* and the *Nairung* on the 10th, at which place orders were received that after landing the native establishments, reserve stores, and forage, the transports were to proceed to Cape Town. October.

By this date matters had reached a climax.

On the 7th October, the reserve for the First Army Corps was called out in Great Britain; on the 9th, President Kruger intimated to Her Majesty's Government that unless all troops on the borders of the Transvaal were withdrawn, all reinforcements to the army in South Africa made during the previous twelve months were re-embarked, and all troops then upon the sea *en route* there were sent back at once; he should, failing a satisfactory answer within forty-eight hours, with great regret be compelled to regard the action of Her Majesty's Government as a formal declaration of war, for the consequences of which he would not hold himself responsible.

On the 10th, Her Majesty's Government replied through Sir Alfred Milner: "You will inform the Government of the South African Republic that the conditions demanded by it are such as we deem it impossible to discuss." So began the South African War. The one side stepped in with apparent impudence, the other with the like indolence.

On the 10th, while proceeding from Durban to

1899. October. Cape Town, the transport *Wardha* encountered a most terrific storm, when 82 horses out of 150, and 7 mules were washed overboard or killed. The following extract of a letter from 2nd Lieut. Lord Frederick Blackwood to his father, gives a most vivid description of the scene on board: "At last the wooden framework of the stabling began to crack, five minutes later one entire side gave way, and two of my chargers, together with four others, were at once washed overboard. Now came the most horrible scene I have ever witnessed: the deck was covered with one struggling mass of horses and mules, mixed up with the broken woodwork, the whole being hurled first to one side of the deck then to the other. All were horribly wounded, most of them with broken legs. My first charger was amongst this lot, but I managed to get a shot at her with my revolver, and so am glad to say that none of my three suffered the torments of the other horses." The captain of the transport, who had been thirty years at sea, said he had never before experienced such a storm. As her rudder had got broken, she made for Port Natal.

The *Nairung* and *Nowshera* reached Cape Town on the 14th and 15th respectively, when the "A" and "D" squadrons at once entrained for De Aar, proceeding later to Orange River Station, where both encamped on the 18th.

Lieut.-Colonel B. Gough, Captain the Hon. C. Willoughby, Captain Lord Douglas Compton, Lieut. and Adjutant E. Bell, and Lieut. V. Brooke rejoined the regiment from leave in England. Captain D. Campbell and Lieut. J. Gordon-Stirling had been sent on from India to Stellenbosch, to organise the regimental transport. 1899. October.

On the 19th, part of the "C" squadron came into camp.

The following is a list of the regimental staff, and squadron officers; together with the senior non-commissioned officers:—

Lieut.-Colonel B. Gough, commanding.	Regimental - Sergeant - Major W. Grant.
Major M. O. Little, second in command.	Rough-Riding-Sergeant-Major C. Wardell.
Lieutenant and Adjutant E. Bell.	Sergeant-Instructor of Fencing A. Beck.
Lieutenant and Quartermaster D. Laing.	Farrier - Quartermaster - Sergeant H. Whittaker.
Maxim gun, Lieut. F. Allhusen.	Orderly - Room Sergeant J. Praeger.

"A" Squadron.	"C" Squadron.	"D" Squadron.
Major Follett.	Captain Lund.	Captain the Hon. C. Willoughby.
Captain Lord D. Compton.	Captain Fiennes.	Captain Ellison.
Captain Campbell.	Lieut. Brooke.	Captain Gordon.
Lieut. Durand.	Lieut. Gordon-Stirling.	Lieut. S. Smyth.
Lieut. Abadie.	2nd Lieut. Lord F. Blackwood.	Lt. the Hon. D. Carleton.
2nd Lieut. Cavendish.	2nd Lieut. Duckett.	Lieut. Theobald.
S.-S.-Major A. Bateman.	S.-S.-Major R. Gidden.	2nd Lt. Sadleir-Jackson.
S.-Qm.-Sergt. D. Ankers.	S.-Qm.-Sergt. L. Bell.	S.-S.-Major A. Pollock.
		S.-Qm.-Sergt. W. Letts.

In locating a ford near Signal Hill, Captain Campbell and Sergeant Westman got into deep water and had a narrow escape from drowning.

On the 23rd, Colonel Scott-Chisholme, who had exchanged some years previously to the 5th Lancers, and was now in command of that fine regiment the Imperial Light Horse; was killed at the head of it at the battle of Elandslaagte.

1899. On the 24th, 2nd Lieut. Brassey arrived from England.

November. On the 9th, "C" and "D" squadrons, with 100 mounted infantry, the regimental maxim gun, and some artillery, all under Major Milton; proceeded on a three days' reconnaisance in the direction of Belmont, and on the 10th came under fire for the first time in the war near Luiperskop. Lieut. V. Brooke had a narrow escape while sketching some ground; his horse was killed and a bullet put through his helmet. He ran for some wire fencing, which delayed the enemy, and was rescued by a patrol under Sergeant Atkins.

On the 12th, Lord Methuen assumed command of the division, and proceeded to organise a column for the relief of Kimberley. The force, about 8000 strong, which gradually assembled at Orange River Station was composed as follows: A brigade of Guards (1st Scots, 3rd Grenadiers, and 1st and 2nd Coldstreams). The other brigade consisted of the 2nd Yorkshire Light Infantry, the 2nd Northamptons, the 1st Northumberlands, and a wing of the North Lancashires, with a Naval Brigade of sailor gunners and marines. The mounted troops consisted of the regiment, some mounted infantry, and the 75th and 18th Batteries Royal Field Artillery.

On the 22nd, a move was made for Kimberley 60 miles distant, and touch with the enemy was obtained at Belmont.

Here on the 23rd, a battle took place. The "D" squadron with Rimington's Guides, under Major Rimington, working eastwards, while the other two squadrons accompanied the main body. The enemy was driven from his position after some very severe hand-to-hand fighting; the loss of the regiment being four men. 1899. November.

On the 24th, an advance was made to reconnoitre the enemy's position at Enslin, and the following morning the "A" and "C" squadrons, with Rimington's Guides and two companies mounted infantry, under Colonel Gough, worked round the left flank of the Boers; while the "D" squadron, one company mounted infantry, and one company New South Wales Lancers, under Major Milton, worked round the right; the enemy's position at the same time being shelled by the Naval Brigade, and Artillery. Major Milton tried to effect a junction between these two forces, but was prevented by the superior numbers of the enemy, and had to retire; which he did safely, much helped by the accurate fire of a party of the New South Wales Lancers. The casualties of the regiment on this occasion were one killed, four wounded, two missing. Captain Campbell had a narrow escape, and was rescued by Lord Douglas Compton. These two officers had gone forward to see if they could make out anything of a force of Boers supposed to be advancing from Ramdam. When they had gone about a mile they saw a man riding towards them, who came within about a hundred yards, then suddenly turned his pony broadside on, slipped over on the side farthest from Campbell, and covered him with his rifle. Campbell galloped off in a slanting direction in order to give the Boer a more difficult shot, but the bullet hit his horse in the head and turned him head over heels, breaking the saddle in his fall. Compton seeing this galloped to him, and Campbell scrambled up behind; but as the horse was

1899. November. very nervous, and showed a strong desire to go for barbed-wire fences, and in addition each thought the other had got the stirrups, the ride was far from pleasant. The Boer fired another shot, which went through Campbell's revolver case.

On the 26th, Lieut. J. Gordon-Stirling, who was out with an early morning patrol, had his left forearm broken by a shot. The patrol was fired upon from a farm where only women were visible, and they invited the party to come up and have some coffee. Stirling was brought back in safety by Corporal Green, who stayed by him under fire.

On the 27th, Major M. O. Little assumed command of the regiment.

On the 28th, the regiment covered the advance of the division, which moved forward to try and effect a crossing of the Modder River, being more or less under fire from 5.15 A.M. until dark. Captain F. F. Colvin arrived from England in time to take part in this day's operations.

December. On the 9th, a reconnaissance in force was made of the enemy's position at Magersfontein, where Cronje was commanding; the troops employed being the regiment, Rimington's Guides, Mounted Infantry, G Battery R.H.A., one Battery R.F.A., and the 4.7 Naval gun "Joe"; the "C" and "D" squadrons covering the advance.

At 3 P.M. on the 10th, the force moved out to take up its position for the attack early next morning. It was raining hard, and the men, with one blanket between two soldiers, bivouacked about three miles from the enemy's position.

About 1 A.M. on the 11th, the final advance was made, and by dawn the battle of Magersfontein was raging. It was an artillery and infantry engagement more particularly, and has been ably described by many writers. 1899. December.

Sergeant Ambrose of the regiment had his horse shot under him, and when it fell pinned him to the ground so that he was unable to extricate himself. In this position he was incessantly fired at by the Boers, who eventually hit him in the arm. He was rescued late at night by some of the Guards Brigade, having suffered intensely from the heat. The casualties of the regiment were 2 men killed, 9 wounded; 8 horses killed, 16 wounded. The force moved back towards Modder River, the retirement being covered by the "C" and "D" squadrons. Captain E. R. Gordon was appointed Brigade-Major to Major-General Babington.

On the 26th, the regiment with 2 guns R.H.A., 50 Mounted Infantry, and a section of Royal Engineers, went up the right bank of the Modder River to destroy a house that was used by the Boers at night, the two advance troops, under Lieutenants Abadie and Durand, coming under fire at 7 A.M.

On the 6th some races were held, and Major Little won the "Modder River Cup" on his bay-waler horse "Orcano," beating a large and varied field of quadrupeds. 1900. January.

On the 15th, the officers played those of the 12th Lancers at hockey, and beat them by 4 goals to 2.

On the 25th, the Cavalry Brigade played the Guards Brigade at hockey, and beat them 4 to love.

1900. January. On the 28th, Lieutenant and Adjutant Bell had to go on the sick list with enteric fever, Captain Colvin taking up his duties.

February. On the 4th, the regiment marched to Koodoosberg.

On the 7th, the regiment had a hard day's work at Paynter's Drift, when Lieutenant Cavendish and 4 men were wounded. The "A" squadron rejoined from duty at Klokfontein.

On the 9th, in accordance with the orders of Lord Roberts, who, with his Chief of the Staff, General Kitchener had recently arrived in the country; a cavalry division was formed under command of Major-General J. D. French, the brigades of which were composed as follows:

1st Brigade.—Brigadier—Colonel Porter, The Carabiniers.
The Carabiniers.
Scots Greys.
The Inniskillings.
14th Hussars.
New South Wales Lancers.
G and T Batteries, R.H.A.

2nd Brigade.—Brigadier—Colonel Broadwood, 12th Lancers.
Household Cavalry Composite Regiment.
10th Hussars.
12th Lancers.
Q and U Batteries, R.H.A.

3rd Brigade.—Brigadier—Colonel Gordon, 15th Hussars.
9th Lancers.
16th Lancers.
O and R Batteries, R.H.A.

The town of Kimberley was at this date still unrelieved, although Lord Methuen's force had marched

in November 1899 from Orange River Station for that 1900.
purpose. It was now deputed to the cavalry division Feb-
to carry out the relief, and the Commander-in-Chief, ruary.
Lord Roberts, thus addressed the commanding officers of regiments composing the division: "The division is about to start on an expedition which it will remember all its life. A British cavalry force of this size has never been employed before, and it rests with the division to show that it can maintain the traditions of the British cavalry and relieve Kimberley at all costs." How well and gallantly it responded to Lord Roberts' appeal history lucidly tells us. An infantry division under General Tucker was detailed to hold the ground made good by the cavalry. Lieut. E. H. Abadie was selected by General French as an A.D.C., and Lieut. F. Allhusen in a similar capacity by General Tucker.

On the 13th, the passage of the Modder River was gained, and the Boers' retreat to Bloemfontein was cut off. The laager of Van de Merves, with stores and ammunition, was taken by the "C" squadron, being well and quickly supported by the "D."

On the 14th, a halt was made at Klip Drift, but early on the 15th the advance was resumed, the division being then about 40 miles from Kimberley. The "C" and "D" squadrons, with two of the 16th Lancers, charged and took two kopjes, from which the Boers had been directing a galling fire on our troops. Sir A. Conan Doyle, writing of this particular incident, says: "Disregarding the Boer fire completely, the cavalry swept in wave after wave over the low nek, and so round the base of the hills. The Boer riflemen must have seen a magnificent military spectacle as regiment

1900. February. after regiment, the 9th Lancers leading, all in very open order, swept across the plain at a gallop, and so passed over the nek." About 2 P.M. the division came in sight of Kimberley, and General French and his staff entered the now relieved town.

The same author makes the following most pertinent remark, which, in the history of a cavalry regiment, may well be reproduced at this point: "The war was a most cruel one for the cavalry, who were handicapped throughout by the nature of the country and the tactics of the enemy. They are certainly the branch of the service which had least opportunity for distinction. The work of scouting and patrolling is the most dangerous duty which a soldier can undertake, and yet from its very nature it can find no chronicler."

These remarks are true and kind to a degree, but I think the general feeling of the regiment is best expressed in the following words: "England expects every man to do his duty." They are brief, to the point, and seem to me to include "all" in the one word "duty." The casualties were 1 killed, 2 wounded; 15 horses killed, 15 wounded.

On the 16th, the 3rd Brigade was ordered to try and cut off a body of the enemy who were trekking along the Leeufontein-Boshof road. The regiment came under very heavy fire, 2nd Lieut. Brassey was killed, Captain Gordon and Lieut. Durand were wounded; and 40 horses were killed or died of exhaustion. It was 9 P.M. before the regiment arrived at Kimberley, having been marching and fighting all day.

The horses were inspected the following day, and it was found there were only 105 fit to go on, out of the 422 which started from Modder River. Major Follett, Captain Ellison, and Lieut. Skeffington-Smyth were left with these at Kimberley. The 2nd Brigade moved off under General French for Koodoesrand Drift to try and intercept General Cronje, who was making a dash for Bloemfontein, being followed early on the 18th, by the 3rd Brigade. The kopjes overlooking the Drift were reached at 5 P.M., and Lieut. Sadleir-Jackson rode forward and reconnoitred the Drift by himself, but as the ground was found unsuitable for a bivouac, the regiment rejoined the Brigade at Blinkwater Pan, having been 18 hours in the saddle and covered more than 40 miles. 1900. February.

General French having decided to occupy the Drifts at Koodoesrand and Makauw, so as to prevent any relief from the south-east; the regiment proceeded to the first-named place, the 16th Lancers to the latter. The 21st was a busy day, all the squadrons being actively engaged, together with two squadrons Household Cavalry and parts of Roberts' and Kitchener's Horse; in clearing the kopjes round Kitchener's Hill south of the Modder River. Captain Campbell and one man were wounded. Lieut. Sadleir-Jackson distinguished himself by his promptitude in seizing a small kopje and driving off about forty of the Boers who were galloping for it. A small party of nine men of the "C" squadron, under Captain Fiennes with Lieut. Lord F. Blackwood, charged into the tail end of some flying Boers, and every man got his lance home.

The strength of the regiment on the 25th, was as follows:—

1900. February.	15 officers.	"A" squadron	60 men,	56 horses.
		"C" „	58 „	44 „
		"D" „	44 „	37 „

On the 27th, General Cronje, who had by this time been completely surrounded in his laager at Paardeberg, surrendered unconditionally to Lord Roberts, and about three thousand of Transvaal troops, and eleven hundred of the Free State were taken prisoners.

March. Early this month a welcome addition of 50 remounts was received.

On the 6th, the cavalry division moved its bivouac to Osfontein.

On the 7th, while executing a turning movement, the "C" squadron came in for some heavy fighting. Lieut. Duckett had his horse shot, and was rescued in a most gallant manner by Corporal Beadle. Other casualties were 2 men and 21 horses killed, 13 men and 11 horses wounded. The men and horses which had been left at Kimberley, after the forced march there, rejoined at this time—105 of the former, 106 of the latter. The 3rd Brigade advanced towards Driefontein, accompanying General Tucker's division. Lieut. Duckett captured a Field-Cornet called Prætorius, who, armed with a rifle and bandolier, said he was out "buck-shooting."

On the 13th, the town of Bloemfontein surrendered to Lord Roberts, who selected the regiment to furnish his escort on entering the capital of the Orange Free State, and in making this selection said: "I know they have always been a splendid regiment, and this is the third time they have been on active service with me. (The Indian Mutiny, Afghan War 1878-1880, South African War)."

Lieut. Sadleir-Jackson's troop led the way, followed by Lord Roberts and his staff, then the "A" and "D" squadrons. The "C" squadron was engaged capturing a German ambulance train, of which the engine succeeded in effecting its escape, although it did so under a heavy and very able fire from the Regimental-Sergeant-Major (W. Grant). 1900. March.

On the 15th, the following appeared in the orders for the Cavalry Division, and as it sums up in a few words the work of that force from the 12th February to this date, is here inserted :—

"On the 12th February this force crossed the boundary which divides the Orange Free State from British territory. Three days later Kimberley was relieved, and on the 15th day (*i.e.* 27th February) the bulk of the Boer army in this State under one of their most trusted Generals (Cronje), was made prisoners of war. Twenty-nine days from the commencement of the operations, the capital of the Orange Free State was occupied.

"This is a record of which any army in the world may be proud, a record which could not have been achieved except by earnest, well-disciplined troops, determined to do their duty and to surmount whatever difficulties or dangers might be encountered. Exposed to extreme heat by day, bivouacking under heavy rain, marching long distances (not infrequently with reduced rations), the endurance, cheerfulness, and gallantry displayed by all ranks are beyond praise, and Lord Roberts feels sure that neither Her Majesty the Queen nor the British nation will be unmindful of the efforts made by this force to uphold the honour of their country."

1900. March. On the 25th, the "C" squadron was actively engaged near Brandfort, Lieut. Duckett, with the advanced troop, getting within 200 yards of the town. The squadron was hard pressed, and had to make a running fight of it for nearly 10 miles, 5 men being wounded and 9 horses killed. Sergeants Andrews and Hart, Corporal Green, and Privates Heanes and Shaw were specially mentioned for their conspicuous gallantry in bringing men out of action.

On the 29th, the cavalry division concentrated, in order to make a turning movement to the west of Brandfort, and get round some kopjes which overlooked that place. The enemy was eventually driven off, and the regiment proceeded to Klipfontein.

April. On the 7th, 168 remounts were received, of which 96 were Argentines, these proved only very moderate mounts. The strength of the regiment was now 367 rank and file, 438 horses.

On the 22nd, the brigade marched for Leeuwkop, and on the 24th the regiment was heavily engaged all day near Roodekop, the "D" squadron suffering severely; 2 men killed, Captain Stanley, Lieut. Brooke, 14 non-commissioned officers and men being badly wounded. This was Captain Stanley's first day under fire, and much to everyone's sorrow his wound proved fatal, and he died on the 3rd May.

On the 28th, a hard day's work was undergone. The regiment left camp at 6 A.M., and did not return till 9 P.M., after fighting all day, when a miserable night was spent. There was no food, forage, or blankets, and it was extremely cold. Lieut. Theobald and part of his patrol was cut off, owing to a squadron of the 17th Lancers which was in support of him, having to retire before a heavy fire. 1900. April.

On the 29th, a fresh bivouac was taken up; the "A" squadron and one of the 17th Lancers being sent out to hold some kopjes to the south-west of Thaba'nchu. The remainder had just settled down to breakfast when a heavy fire was heard from the direction of the outposts. The squadron of the 17th Lancers had retired, losing its Maxim gun; and the "A" squadron severely pressed on both flanks, was obliged to do the same; the retirement being carried out with great regularity. While the fighting was going on, Lord D. Compton had sent back two men to the led horses for more ammunition; when they returned to the kopjes they found the squadron had gone. Sergeant Mercer volunteered to go out and try to find them; this he did, successfully brought them down under a heavy fire, and was recommended for the Victoria Cross.

On the 30th, the 3rd Brigade bivouacked at Israel's Poort.

The "A" squadron at this time formed part of a composite regiment sent out to aid General Ian Hamilton. May.

On the 3rd, Captain Stanley succumbed to his wounds received on the 24th April, and was buried by the Coldstream Guards at Bloemfontein.

On the 5th, a welcome addition of 127 remounts was received.

On the 9th, the brigade joined the main army under Lord Roberts at Welgelegen, and was appointed as

1900. May. divisional troops; the strength of the regiment being 21 officers, 384 men, 374 horses.

On the 29th, the army reached Rietfontein, 8 miles east of Johannesburg.

June. On the 2nd, a squadron of 88 men and 98 horses, in command of Captain the Hon. C. Willoughby, together with Captain Campbell, Lieutenants Durand, Lord F. Blackwood, Pollok, and Sadleir-Jackson, moved out with a composite squadron of the 16th and 17th Lancers, the whole under Major Hunter-Weston, R.E.; to endeavour to blow up the railway bridge at Bronkhorst Spruit, on the Pretoria-Delagoa line. Burnham, the American scout, accompanied the force. It was 65 miles to the bridge, which had to be destroyed by 2 P.M. on the 3rd.

The troops set off about 8 P.M., the 16th and 17th Lancers in front. They moved in column of files, leading the horses. So, all night with occasional halts, the march proceeded, Burnham making good the ground in advance. About 6 A.M. the force mounted, and had hardly done so before two or three Boers showed themselves, and Burnham reported three large laagers right ahead. The situation was awkward. A force about 250 strong, 25 miles from any support, was confronted by one not less than 1000. Wise discretion, unfortunately too seldom exercised in the campaign, counselled a retirement. This was executed in a most methodical and orderly manner. The regiment took the left, the 16th and 17th the right, both retiring simultaneously by succession of troops. The enemy pressed them very hard, and it was only by the skilful way in which troop leaders handled their troops that

disaster was averted. The retirement continued in this 1900.
way for about ten miles, when it was found possible June.
to take up a defensive position near Bapsfontein, which the enemy disliked the look of and drew off.

It was a costly day's work for the regiment. Lieut. Pollok was mortally wounded and died at Boksberg; Lieutenants Durand (who eventually nearly had to have his leg amputated on account of blood-poisoning) and Sadleir-Jackson were badly wounded, but continued to lead their troops in the retirement; Lord F. Blackwood and Captain Campbell had their horses killed under them. The former was rescued by Privates Wright and Albert, both of whom were honourably mentioned. Other casualties, were 1 man and 16 horses killed, 4 men and 5 horses wounded.

On the 5th, Lord Roberts with the main body of the army entered Pretoria. A man of the regiment, who had carried with him all through the previous operations a lance flag, for use on this occasion, was by Colonel Little appointed his orderly. Much to his disappointment he was unable to make use of it, as when the troops were concentrating for entry into the town, the regiment was suddenly detailed to clear the country on the south-east of it. Lieut. Theobald, who had been taken prisoner on the 28th April, and had been in the gaol here, appeared again. (For full particulars of his experiences see Lieut.-Colonel F. Colvin's and Captain E. R. Gordon's 'Diary of the War.')

On the 7th, the regiment lost yet another good officer, Captain R. Ellison dying of fever at Kroonstad.

1900. June. On the 8th, the 3rd and the 2nd Brigades were together at Swavel's Poort. Having reached Pretoria the capital of the Transvaal, Lord Roberts issued a congratulatory Order to the Army, from which the following paragraph is of interest:—

No. 5. The cheerful spirit with which difficulties have been overcome and hardships disregarded are deserving of the highest praise; and in thanking all ranks for their successful efforts to obtain the object in view, Lord Roberts is proud to think that the soldiers under his command have worthily upheld the traditions of Her Majesty's army in fighting, in marching, and in the admirable discipline which has been maintained throughout a period of no ordinary trial and difficulty.

The strength of the three squadrons being now so very small—

"A"	87 men	55 horses
"C"	85 "	47 "
"D"	90 "	57 "
	262	159

the regiment was formed into two squadrons, under Captains The Hon. C. Willoughby and F. Lund. The 3rd Brigade was attached to General Ian Hamilton's division, and in conjunction with the 2nd, protected the right flank and rear while the division was fighting the action of Diamond Hill.

On the 15th, the regiment started back to Pretoria.

On the 17th, Lord C. Bentinck, who had been on duty with General Baden-Powell during the siege of Mafeking, rejoined the regiment.

On the 19th, the 2nd and 3rd Brigades moved with General Ian Hamilton's division towards Heidelberg. 1900. June.

On the 22nd, Lieutenants Sadleir-Jackson and Trower captured a party of 16 Boers, among them being a Field-Cornet. Lieut. Duckett was appointed Provost-Marshal to the 3rd Brigade.

On the 27th, the 2nd and 3rd Brigades left Heidelberg to form a sweeping force on the east, while General Bruce Hamilton's troops did the same on the west.

On the 29th, the brigades crossed the Vaal River and re-entered the Orange River Colony (late the Orange Free State).

On the 3rd, 114 men and 140 horses joined from Kroonstad, the strength of the regiment at this date being 17 officers, 291 non-commissioned officers and men, and 313 horses. July.

On the 15th, the regiment was at Heilbron. Colonel Little was appointed to command the 3rd Brigade in place of Colonel Gordon, who was moved to command the 1st Brigade in place of Colonel Porter, who had been wounded. The command of the regiment then devolved on Major F. Colvin.

On the 19th, the regiment had a sharp afternoon's fighting with De Wet's troops near a farm called Paardeplaats, 2 men being wounded.

On the 23rd, Captain the Hon. Claude Willoughby had to go on the sick list, and as the regiment was very weak in numbers (205 men, 177 horses), his squadron ("D") was broken up.

On the 31st, the "A" squadron under Lord Douglas Compton, was engaged on a reconnaissance

1900. July. towards Rhenoster, and had some sharp work. A small patrol under Lieut. Wynn was nearly cut off, they had to gallop through the enemy, and Sergeant-Major Ankers behaved in a plucky way. Seeing Private Daniel's horse was shot, he slipped off his, and covered him until Daniel could get back to him. When he did, they both succeeded in getting safely back to the squadron on foot.

August. On the 4th, the brigade was employed supporting General Knox's advance on Rhebokfontein, joining the 2nd Brigade on the 8th, at Parys.

On the 12th, the regiment acting as support to Lord Methuen's force which was in pursuit of De Wet, bivouacked at Schoolplats. A gun, ammunition, and some of his prisoners were captured, among them being Privates Albert and Lanham, who had had rather a bad time of it.

On the 14th, Lord Methuen's camp at Rietfontein was reached.

On the 15th, the 2nd and 3rd Brigades left for the relief of Colonel Hore, who, with about 500 Australian troops—Victorians, New South Welshmen, Queenslanders—and some of the Rhodesian regiment, was besieged by Delarey at Brakfontein, on the Elands River. This garrison had been escorting a large convoy, and had fortunately taken up a good position before being surrounded. Delarey had about 2500 men and six guns, which commanded the place from all sides. The garrison held out for eleven days, during which time it is computed that 1800 shells were fired into their lines. They had however, with extraordinary energy and skill, dug themselves into shelters, so that they escaped annihilation. They lost nearly all their horses and oxen, and 75 men were killed or wounded. The relief came at last, and the gallant resistance of this small force stands out as one of the finest performances of the war. **1900. August.**

On the 25th, the regiment started for Pretoria with 16 officers, 153 men, 109 horses; the other regiments of the brigade being in the same emaciated condition. The enemy was encountered after going a few miles (near Botha's Farm). The regiment acted dismounted, Brigadier Little receiving a very nasty flesh wound high up on the thigh, and the command of the brigade then devolved on Colonel Dalgety.

On the 31st, the left of the advance became engaged near Quaggafontein. The "A" squadron under Lord D. Compton, and the Maxim gun under Captain Gordon, were employed on that flank, the "C" squadron being on the right. Hot firing took place, but as no impression could be made on the enemy, the order was given to hold on to the position and retire at dusk. The casualties in the force were 50; Lieut. Skeffington-Smyth behaved with great gallantry assisting the wounded, and was himself hit in three places late in the day.

The regiment proceeded to Kroonstad, there to refit. The strength being 15 officers, 150 men, 93 horses, of which latter 35 only were fit for work. A welcome addition of 185 horses was here received. **September.**

On the 22nd, a polo tournament was held.

FIRST TIES.

The Yeomanry beat The Garrison.
First team 17th Lancers „ 9th Lancers.

1900. September.

SECOND TIES.

Second team 17th Lancers beat Yeomanry.

FINAL.

First team 17th Lancers beat Second team 17th Lancers.

Captain Fiennes, Lieutenants Lord F. Blackwood, Cavendish, and Duckett represented the regiment. The regiment now, with 15 officers, 301 men, and 287 horses, was formed into three squadrons.

"A" . . .	Captain Lord Douglas Compton.
"C" . . .	,, F. Lund.
"D" . . .	,, E. R. Gordon.

On the 23rd, the brigade left Kroonstad on a rounding-up expedition, and on reaching Vecht Kop (which is just over the river Rhenoster), the regiment, which was covering the advance became engaged, and so continued until Rietfontein was reached, where the force bivouacked; the "A" squadron under Lieut. Wynn, furnishing the outposts. The camp had hardly settled down, when shells from at least 7000 yards off came hurtling into it. As the outposts were hard pressed, the brigade went to their support, the substitute regiment being the first under arms; as soon however, as our guns got into action the enemy drew off. The brigadier expressed himself as highly pleased with the quick way in which the regiment had turned out, as also with Lieut. Wynn and his party, for holding his position so successfully.

October. On the 13th, the strength of the regiment was as follows: 16 officers, 337 men, 316 horses.

On the 16th, the "A" squadron had a brush with

the enemy, the brigade reaching Rietgat. Between this date and the end of the month the regiment was several times engaged with the enemy, the last time at Ventersburg, when the Boers declined to come to close quarters. 1900. October.

On the 26th, Major Follett rejoined and assumed command.

On the 2nd, Lieut. Theobald once more rejoined. After being liberated from Pretoria he was detailed for duty with a composite squadron under General French's command, and while so serving got enteric fever. November.

On the 10th, the 3rd Brigade which latterly had consisted of the 9th, 16th, and 17th Lancers under Brigadier Porter, reached Bloemfontein; there an order from Lord Roberts was received that it was to be broken up, and three mobile columns of all arms under General Charles Knox were to be formed. A day or two later about sixty officers of the brigade dined together at the Bloemfontein Club, the guest of the evening being Brigadier Porter. In accordance with the above order "A" and "D" squadrons were posted to Colonel Barker's column, which consisted of the following troops: Lovat's Scouts, two guns, one pom-pom, two companies each the Dublin Fusiliers and Seaforth Highlanders. "C" squadron with the 17th Lancers was posted to Colonel Herbert's column. The former was to work westward from Springfontein, the latter eastward from Edenburg. The lances and the carbines were returned into store, and the regiment for the future carried the rifle; returning to the manner of arming of the year 1715. The two columns were

1900. November. together at Redderburg on the 22nd, and on the three following days were both actively engaged.

On the 27th, a column under Colonel Pilcher joined the force, and all bivouacked at Helvetia.

December. On the 2nd, Colonel Barker's column left for Slik Spruit, and about 11 A.M. heard Colonel Herbert's column heavily engaged. Lieuts. Sadleir-Jackson's and Theobald's troops were in the thick of it, the latter being most severely wounded in the thigh as well as having his hand grazed by a bullet. Captain Forrest, R.A.M.C. (the regimental surgeon), tried to get him into a safe place, but was prevented by a galling fire from the Boers. Privates Lentchford and Vincent also undertook the task, but were unable to carry out his removal. Later in the day an ambulance was allowed to come up and he was removed. No food or blankets reached the force till 11 P.M., and the men bivouacked on the ground they stood on.

On the 3d, the squadrons were fighting all day. Lieut. Gordon-Stirling and four men were wounded.

Up to the 14th, the columns had been hotly pursuing De Wet who had had to turn northwards, but on this day it was hoped that at last he was cornered. A helio message was received from Thaba'nchu Hill about 10 A.M. saying that he had tried to force his way through Springhaan Nek but had failed. In reality this was only a reconnaissance, and when he found that the nek was not very strongly held, he decided to make a dash for it. As soon as the message came in, off went the columns; the bad weather, the long marches, the short rations of the last three weeks were all forgotten. They were however, too late, De Wet had quickly made up his mind, and galloped at full speed in open order through the nek, braving the long-range fire of rifles and guns, and so vanished into the mountainous country around Ficksburg. So hurried was his flight that he left some waggons, ambulances, small-arm ammunition, a pom-pom, and one of our guns belonging to the 68th Battery. 1900. December.

On the 24th, Lieut. Lord F. Blackwood, who was with the "C" squadron near Burghersdorp, was severely wounded. The regiment had some hard and long days' work during the last few days of the year, and on the 1st January bivouacked at Lindley, where also were Colonels White's and Pilcher's columns.

On the 13th, the regiment was detailed to act with Colonel White's column. 1901. January.

On the 17th, a very sharp rear-guard action was fought on the march to Dorn Kop, the troops under Lieut. Sadleir-Jackson and Sergeant Draper having a warm time of it.

On the 23rd, the regiment was again moved to another column, that of Colonel Maxwell.

On the 3rd, three columns commanded by Colonels Maxwell, White, and Munro, all under General Bruce Hamilton, assembled at Bethulie for assortment. Major F. Colvin was selected to act as staff officer to another column under Colonel Crewe. February.

From the 10th to the 26th, the column of which the regiment formed part was acting on the defensive to prevent De Wet from coming south. They had some very trying marches owing to difficult ground,

1901. scarcity of provisions, and forage, and also encountered some fearful storms.

March. On the 7th, Aliwal North was reached by train, after a long and tiresome journey; and on the 10th, Bruce Hamilton's columns commenced a systematic clearing of the country of all inhabitants and stock.

On the 29th, the column to which the regiment belonged was detailed to escort the wrecklings of the three columns to Edenburg. These consisted of the following: 33,000 sheep, 1000 head of cattle, some ponies, 1100 white and 900 black refugees. It was by no means a pleasant duty, and the rear-guard had an anxious time of it. Eventually Bethanie was safely reached on the evening of the 31st.

April. On the 3rd, the "C" squadron once more rejoined the regiment at Springfontein, and Sergeant-Major Wardell, who had been appointed regimental-sergeant-major in place of W. Grant (promoted), took up his duties as such.

Major F. Colvin was given command of a column in Brigadier-General Plumer's force.

During this month farm-clearing operations went on, continual brushes with the enemy taking place. A disastrous one occurred on the 17th. Two troops of "D" and two of "C" squadrons were out; one of the latter under 2nd Lieut. the Hon. A. Macdonald, fell into a trap. A party of Boers had hidden themselves among some kopjes, and sent a few men to canter across their front as a bait. Macdonald gave chase and was fairly caught, but he refused to surrender, and together with four men lost his life, seven others being wounded. With reference to this most lamentable occurrence the following order was on 16th May published by General Bruce Hamilton: "The Major-General Commanding much regrets that he has to record the death of 2nd Lieut. the Hon. A. Macdonald and four men of the 9th Lancers, who were killed in action at Werfkrall on the 17th April. 2nd Lieut. Macdonald, with his troop of only twelve men, galloped for a kopje held by the Boers, who were hidden behind sangars. They were outnumbered by the enemy, but refused to surrender, and held their ground until, without exception, every man was hit. The whole party behaved in a most gallant manner, and the Major-General Commanding deeply deplores their loss." 1901. April.

On the 18th, the column commander, Colonel Maxwell, met with an accident. The late severe and very cold weather had tried him severely, and while watching his force cross a donga he fainted, and his pony fell back on to him. He was taken to Rouxville, but died there on the 20th. Although in delicate health he never flinched from any hardship, and his death was a sad loss to the column. May.

On the 27th, "A" and "D" squadrons were farm-clearing, and as the "A" would eventually have to get over the nek in the hills overlooking Ospoort, the emergency squadron "C" was sent in support. Lieut. Gordon-Stirling was sent with his troop, covered by the pom-pom, to occupy the nek. A party of Boers was concealed, and allowed the advanced points to get right through the nek, and the troop under the kopje; when the order was given of "Hands up." Lieut. Gordon-Stirling ordered files about at once, and the men took what cover they could. Reinforcements soon

1901. May. arrived, and the Boers were obliged to withdraw. Lieut. Gordon-Stirling went under a heavy fire to help Sergeant Sharples, who had had his horse shot. Our casualties were — killed and died of wounds, 5; wounded, 3.

June. On the 2nd, the regiment reached Dordrecht, having come by train from Edenburg, and on the following day was ordered to form part of Colonel Scobell's column, in which some old friends, the Cape Mounted Rifles, were serving.

On the 6th, the column made a night march starting at 9 P.M., and at dawn was actively engaged near Myburg's Farm on the Kraai River. The enemy was driven off, and saddles, ammunition, &c., together with 163 ponies were captured.

On the 12th, General French assumed command of all the operations in Cape Colony. Captain Lund was given command of a column of Yeomanry, and was permitted to take his own squadron "C" with him, as part of his force. He proceeded to De Aar.

On the 25th, the "A" and "D" squadrons reached Graff Reinet, after a very bad and trying railway journey of three days, in order to refit. The men had no food for twenty-four hours, and the horses only forage on two days.

July. On the 9th, a skittle gymkana was held at Graff Reinet, when there was a great gathering of all the rank, fashion, and beauty there.

On the 10th, the columns commanded by Scobell, Crewe, Windham, and Doran were ordered by General French to make a combined movement to try and catch Scheepers, a young but very energetic and resourceful leader, who was hiding in the hills about thirty miles west of Graff Reinet. 1901. July.

Scobell's column was at Elands Poort by dawn of the 12th. The day was spent hunting the enemy about, and eventually a chain of small posts about 200 yards apart was formed. Scheepers however, was not to be caught, and he made his escape with the bulk of his commando through a long kloof, up the precipitous side of a hill. Lieut. Theobald captured a Boer who was carrying despatches to Scheepers. Lord D. Compton and Lieut. Bell captured several Boers, among whom was Lieut. Limburgher. Lieut. Wynn had a narrow escape, a bullet going through the brim of his hat. Captain Bell was appointed staff officer to Colonel Hunter-Weston.

On the 28th, General French inspected the regiment, and in his remarks spoke specially of the gallop of the 3rd Cavalry Brigade (9th and 16th Lancers) from Klip Drift to Kimberley, and said that when he gave the order "he was afraid there was a chance of but few returning. However, it was successfully carried out, and enabled his cavalry division to perform the magnificent feat of relieving Kimberley, which was the making of his own reputation."

The troops concentrated at Visch River Station on the 8th, and at dawn on the 9th encountered the enemy, when some severe fighting took place, during which Lieut. Theobald was badly wounded in the stomach; 1 man killed, and 2 wounded. Camp was not reached till past midnight, the regiment having been in the saddle twenty-two hours, and having covered more than fifty miles. Lieut. Theobald (one more loss of a most August.

1901. August. promising officer) succumbed to his wounds on the 11th, and was buried at Wildehorndeberg on the 12th.

September. On the 4th, the column bivouacked in the Petrusburg valley, when it was reported that the Boers under Lotter were only nine miles away to the south.

The force started at 1 A.M. on the 5th—"D" squadron on the right, the Cape Mounted Rifles in the centre, and "A" squadron on the left. The commando was found in a kraal near Bouwer's Hork, to the west of Cradock, and so safe did they think their position that they had dispensed with outposts. After some severe fighting, the whole commando with all that it possessed was in Colonel Scobell's hands,—a good reward for the previous week's hard work. The casualties of the regiment were 7 men killed and 5 wounded. In a Column Order of this date, the commanding officer said: "Colonel Scobell's thanks are particularly due to Captain Lord D. Compton and the "A" Squadron, 9th Lancers, for the splendid dash and bravery they showed to-day, and he has already brought the names of the Officers of this Squadron to the notice of the General Officer Commanding."

The column returned to Graff Reinet with the prisoners on the 7th, and remained there a week to rest and refit. The boots of the men were in a very bad state, all the hill climbing they had recently done having completely worn them out.

October. On the 30th, orders were received to march to Middleton and there entrain, which, after many most vexatious delays, was completed by the 2nd October.

On the 13th, a Boer dressed in 17th Lancers' uniform was captured. He was tried by court-martial and shot. 1901. October.

Since leaving Graff Reinet on the 14th September the following distances were covered, which, considering the extremely rough and difficult nature of the country traversed, coupled with the scarcity of food and water, must be considered very satisfactory work—

14th to 30th September .	251 miles.
6th to 20th October . .	343 miles.

On the 2nd of this month, the column was at Beaufort West. November.

On the 17th, Lovat's Scouts, a squadron 17th Lancers, and a Tasmanian contingent joined Colonel Scobell's column, which kept on the move till the end of the month; the "D" squadron and Cape Mounted Rifles under Major Follett working south-west, while the rest of the column under Colonel Scobell took the east and north.

Between the 14th and 26th, the column covered 235 miles, of which the greater part was done by night.

The "A" squadron under Lord D. Compton, which had been detached since the 25th November, joined the column again at Modder Poort on the 1st. December.

On the 20th, the squadrons were once more together at Dordrecht. The last fortnight's work had been very hard on men and horses; the country had been very hilly and difficult, the weather extremely bad, and forage and rations rather short.

1901. December. On the 27th, much to the regret of all, Colonel Scobell was obliged, on account of ill health, to give up his command and go home on sick-leave. Major Follett took over command of his column, with the exception of the Tasmanian contingent.

1902. January. Early in this month the column was at Barkly East, and on the 15th, our old friends the Cape Mounted Rifles, left the column.

February. On the 14th, Three Sisters was reached, and the regiment was ordered to hand over 200 horses to the 5th Lancers. When however a comparison of the horses was made, theirs were found to be in better condition than those of the 9th, so the order was countermanded. On the 25th, both squadrons were heavily engaged near Hout Kraal, as also on the two following days.

March. On the 3rd, the regiment was in action for the last time in South Africa, having a long and varied day's work.

On the 4th, orders were received to proceed to Victoria Road Station and there demobilize.

On the 10th, the "C" squadron rejoined, and the following day the three squadrons entrained for Cape Town which was reached on the 13th.

The following is a table of the distances covered during the campaign, which does not include the ground covered when fighting, or on independent reconnaissances :—

1902. March.

Month.	1899.	1900.	1901.	1902.
	Miles.	Miles.	Miles.	Miles.
January .	...	118	388	389
February .	...	194	397	428
March .	...	178	277	128
April . .	...	152	393	...
May . .	...	313	275	...
June . .	...	201	248	...
July . .	...	343	310	...
August .	...	363	481	...
September .	...	159	366	...
October .	...	303	464	...
November .	141	314	347	...
December .	...	427	433	...
Totals .	141	3065	4379	945

Grand total . . . 8530 miles.

The regiment served in South Africa from the 14th October 1899 to March 1902, and was present at the following engagements and operations :—

The advance on Kimberley, including the actions at Belmont (23rd November 1899), Enslin (25th November 1899), Modder River (28th November 1899), Magersfontein (11th December 1899), and the Relief of Kimberley.

The operations in the Orange Free State from February to May 1900, including the investment of and surrender at Paardeberg (17th to 26th February 1900); the action at Poplar Grove on the 7th March; Karee Siding (29th March 1900); Hout Nek or Thaba'nchu (1st May 1900); and Zand River (10th May 1900).

The operations in the Transvaal State, May and June 1900. —Action near Johannesburg (29th May); entry of Pretoria (4th June); Diamond Hill (11th and 12th June).

The operations west of Pretoria, from July to 29th November, including the action at Elands River.

The operations in the Orange River Colony, May to

1902. March. November 1900, including the engagements at Wittebergen (1st to 8th July), and Caledon River (27th to 29th November).

The clasps awarded to the regiment as a whole are as under:—to the Queen's Medal (South Africa)—"Belmont," "Modder River," "Relief of Kimberley," "Paardeberg," "Johannesberg," "Diamond Hill," "Wittebergen"; the King's Medal, 1901-1902.

Note.—In September 1904, the honours for the South African War which terminated 31st May 1902, have not been officially announced.

The following officers, non-commissioned officers, and men were mentioned in despatches at various times:—

Brevet-Colonel M. O. Little—three times.
Major Follett.
" F. Colvin—twice.
Captain Lord D. Compton.
" F. Lund.
" E. R. Gordon.
" E. F. Bell.
" Lord C. Bentinck.
Lieut. F. H. Allhusen.
" Victor Brooke—twice.
Lieut. J. Gordon-Stirling—twice.
" G. Skeffington-Smyth.
" E. H. Abadie.
" Lord F. Blackwood—twice.
2nd Lieut. L. W. Sadleir-Jackson—three times.
Lieut. S. Theobald.
" D. K. Tooth.
" R. V. Wynn.
2nd Lieut. G. M. Neilson.

Regimental-Sergeant-Major W. Grant (now Quartermaster).
Squadron-Sergeant-Major G. Drury.
" " R. Gidden.
Rough-Riding-Sergeant-Major C. Wardell (now Regimental-Sergeant-Major).
Squadron-Quartermaster-Sergeant D. Ankers.
" " " L. C. Bell.

1902. March.

Sergeant	F. Andrews—twice.	Private	J. Bradfield.
"	F. Beckett.	"	R. Belsey.
"	G. Casebow.	"	Capon.
"	H. Cook.	"	W. Lentchford.
"	R. J. Mason—twice.	"	T. Holman.
"	J. Mercer.	"	C. Stanford.
"	T. M. Smith.	"	S. Taylor.
Corporal	C. Green.	"	S. Vincent—twice.
"	T. Mitchell.	"	J. Whitton.
"	Pearson.	"	Wilcox.
"	C. Wilson.	Trumpeter	J. Pelter.
Private	S. Albert.		

The following officers were awarded the Order for Distinguished Service :—

Lieut.	Lord F. Blackwood.	Lieut.	Victor Brooke.
"	L. Sadleir-Jackson.	"	E. H. Abadie.
"	G. Skeffington-Smyth.	"	J. Gordon-Stirling.

Brevet promotion was also granted to the following :—

Lieut.-Colonel M. O. Little, to Colonel.
Major F. Colvin, to Lieut.-Colonel.
" S. W. Follett, to Lieut.-Colonel.
Captain F. T. Lund, to Major, and Lieut.-Colonel.
" Lord D. Compton, to Major.
" Lord C. Bentinck, " "
" D. Campbell, " "

The Distinguished Conduct Medal was awarded to the following warrant officer, non-commissioned officers, and man :—

Regimental-Sergeant-Major W. Grant.
Squadron-Sergeant-Major R. Gidden.
" " D. Drury.

1902. March.

Sergeant	F. Andrews—with clasp.	Sergeant	R. Mason.
,,	F. Beckett.	,,	J. Mercer.
,,	H. Cook.	Private	S. Albert.

The Commander-in-Chief specially promoted Corporal Pearson to the rank of Sergeant, and Privates Wilcox and Capon to be Corporals.

I have left the saddest part of the war's mark on the history of the regiment until the end; and all those who are, and have been connected with it, will read the following tables with deep sorrow and regret. They record the loss of relations, best of friends, and most gallant of comrades. Perhaps our one consolation may be that they all died or were wounded doing their duty to Queen, King, and country; and upholding the honourable traditions of the most honourable, well-disciplined, and whole-hearted regiment in the service.

Without undue pride on our own behalf, I should like to emphasise the fact that, throughout the foregoing and following pages, praise has always been awarded to the regiment for its good discipline, good-fellowship, steadiness, and reliability under all circumstances; in fact, generally speaking, for its example as what a cavalry regiment should be. That these traditions may be upheld and stoutly maintained in the future, is in a great measure the reason for these pages having been written: to remind those of us now living what has been done, and to perpetuate that memory for those to come.

1902. March.

CASUALTY LIST.

Killed or died of wounds .	5 officers	40 men.
Wounded	15 „	115 „
Died of disease or accident .	1 „	25 „
Prisoners	1 „	23 „

OFFICERS KILLED.

2nd Lt. P. F. Brassey . . .	Dronfield	16th Feb. 1900.
„ The Hon. A. Macdonald	Vaalkop	17th April 1901.

DIED OF WOUNDS.

Captain H. Stanley	Rooikop	3rd May 1900.
Lieut. S. R. Theobald	Wolve Vlei	11th Aug. 1901.
2nd Lieut. J. Pollok	Orange Grove	June 1900.

DIED OF DISEASE.

Captain R. Ellison	Kroonstad	June 1900.

The following officers were wounded :—

Lieut.-Colonel M. O. Little	at	Zeerust.
Captain E. Gordon	„	Dronfield.
„ D. Campbell	„	Paardeberg.
„ Lord C. Bentinck	„	Mafeking.
Lieut. V. Brooke	„	Rooikop.
„ G. Skeffington-Smyth	„	Quaggafontein.
„ J. Gordon-Stirling	„	Enslin and Slik Spruit.
„ H. Durand	„	Dronfield and Orange Grove.
„ Lord F. Blackwood	near	Burghersdorp.
„ F. Cavendish	at	Koodoosberg Drift.
2nd Lieut. S. Theobald	„	Slik Spruit.
„ L. Sadleir-Jackson	„	Orange Grove.
„ R. Chadwick	„	Leeuwfontein.

2nd Lieut. S. Theobald was taken prisoner near Thaba'nchu.

1902. March. On the 15th, in St. James's Church, Graff Reinet, General French unveiled two brass tablets; one to the memory of men of the 9th Lancers and Cape Mounted Rifles, the other to those of the 2nd Battalion Coldstream Guards, who had lost their lives in South Africa. The dedication service was conducted by the Rev. J. H. Carter, the rector of St. James's.

The tablet on the South Wall is engraved as follows:—

"Dulce et decorum est, pro patria mori."

To the glory of God,
And in Memory of the 9TH LANCERS.

Sergeant John Brighton Mercer.
Lance-Corporal William John Priest.
Private William Vipond.
,, Henry Lionel Clifford.
,, William Kennedy.
,, Victor Ward.
,, Augustus Dodge.

CAPE MOUNTED RIFLES.

Corporal Waldeman Reder.
Private Ronald R. Garthorne.

On the corners of the tablet in silver relief, are the badges of the 9th Lancers and the Cape Mounted Rifles. That of the Coldstream Guards is on the North Wall.

On the 16th, the regiment embarked on board the *Mohawk*, and sailed at mid-day for Durban. Before leaving Cape Town, the following telegram was received from General French :—" I deeply regret that operations which require my urgent attention prevent my going personally to say good-bye to the 9th Lancers, and to congratulate them upon all the splendid work they have done during the last two and a half years. During the greater part of that time it has been my good fortune to be associated with you, and to have witnessed on many occasions how splendidly have all maintained the traditions of your magnificent regiment throughout the war. I deeply mourn the exceptionally heavy losses which you have sustained, and my earnest and heart-felt good wishes go with you, as also my grateful thanks for your valuable help on many trying occasions." 1902. March.

Durban was reached on the 19th, and after taking the 5th Dragoon Guards on board, the *Mohawk* sailed for Bombay on the 23rd. Prior to leaving Durban the following message was received from Lord Kitchener, Commander-in-Chief in South Africa :—" Please tell all ranks how sorry I am to lose their valuable services, but after close on two and a half years' active operations in South Africa I feel that you have well deserved a change. In wishing you good-bye and good luck, I feel I am only giving expression to the wishes of the army in South Africa."

The following officers were left in South Africa, being employed on the staff there :—Colonel M. O. Little ; Brevet-Lieut.-Colonel F. Colvin ; Captains D. Campbell and V. Brooke ; Lieutenants T. Kincaid-Smith and E. H. Abadie.

April. After an uneventful voyage Bombay was reached on the 5th, the regiment entrained the same day, and

1902. April. arrived at Sialkote on the 9th, being joined there by the reserve squadron, which had remained at Muttra during the war.

The following officers landed in India with the regiment:—Majors Follett, Lund, and Lord Douglas Compton; Captains Gordon and Bell; Lieutenants Gordon-Stirling, Durand, Cavendish, Sadleir-Jackson, Wynn, Trower, and Tooth; 2nd Lieutenants Neilson, Brocklebank, Wood, the Earl of Rocksavage, the Hon. R. Cole; and Quartermaster D. Laing. It was at once organised into four service squadrons, and rapidly made up to the Indian establishment.

May. On the 22nd, Lieut. and Adjutant J. Gordon-Stirling, D.S.O., who had done so exceedingly well on all occasions throughout the war, died at Sialkote, the result of an accident when playing polo.

August. On the 15th, Lieut. and Riding-Master W. Baxter died of enteric fever at Sialkote.

October. On the 16th, the regiment marched to Umballa in order to take part in manœuvres to be held between that cantonment and Delhi; these took place between December. the 28th November and the 21st December, on which date it encamped at Budlee-ki-Serai, where, during the early part of the Indian Mutiny, it had rendered such excellent service.

1903. From the 21st December 1902 to the 13th January, it took part in the Coronation Durbar held at Delhi by the Viceroy, to proclaim King Edward VII. Emperor of India; and was detailed as escort for H.R.H. Field-Marshal the Duke of Connaught.

March. On the latter date it marched from Delhi to Rawal-

Pindi, which was reached on the 1st March, the depôt 1903. from Sialkote having already moved there.

By Army Order $\frac{39}{1903}$ the regiment was ordered to August. be armed with the magazine Lee-Metford rifle. The use of the lance for drill, at manœuvres, and for guards was abolished.

A revised edition of the Standing Orders of the Regiment was issued by Colonel M. O. Little. Forty-three years previously his father did the same.

APPENDIX I.

BATTLE ROLL.

PENINSULA.

BY letter dated the 29th March 1815, the Prince Regent was pleased to direct that in commemoration of the services of the regiment in Spain and Portugal, it should bear on its colours and appointments the word "Peninsula." (See medal.)

PUNNIAR.

By General Order in India dated 21st August 1844, the regiment was permitted to wear on its colours and appointments the word "Punniar."

A bronze star made from the guns captured in this action, was presented by the Government of India to the officers and men of the Royal Services engaged therein. (See medal.)

SABRAON.

Her Majesty the Queen, by letter dated 8th June 1847, was pleased to direct that the regiment should bear on its colours and appointments the word "Sabraon."

By order of the Governor-General of India, with the concurrence of her Majesty the Queen, a medal was granted to officers and men of her Majesty's and the East India Company's armies who were engaged in any of the following battles: "Moodkee," "Ferozeshuhur," "Aliwal," "Sabraon." (See medal.)

PUNJAUB, CHILLIANWALLAH, GOOJERAT.

By General Order in India dated 9th March 1853, the regiment was permitted to bear on its colours and appointments the words "Punjaub," "Chillianwallah," and "Goojerat." (See medal.)

DELHI, LUCKNOW.

By General Order 834, Horse Guards, 3rd September 1863, the regiment was authorised to wear upon its colours and appointments the above words, in commemoration of its service at those places. (See medal.)

Note.—It is believed the 9th Lancers was the only regiment at the three main achievements of the Indian Mutiny—viz., "The Siege of Delhi," "The Siege" and "The Relief" of Lucknow. (See medal.)

CHARASIAH ; KABUL, 1879 ; KANDAHAR, 1880 ; AFGHANISTAN, 1878-1880.

By General Order 56, 1881, the regiment was authorised to wear upon its colours and appointments the above words. (See medal.)

SOUTH AFRICAN HONOURS.

In September 1904, these are still not officially announced, but it it believed they will be as under :—

Operations in the "Orange Free State," the "Transvaal," and the "Orange River Colony."

Clasps to H.M. Queen Victoria's South African Medal: for Belmont, Modder River, Relief of Kimberley, Paardeberg, Johannesberg, Diamond Hill, Wittebergen.

H.M. King Edward VII.'s Medal, 1901-1902.

APPENDIX II.

STANDARDS AND GUIDONS.

THE history of standards and guidons as borne in the cavalry is of some interest, and before describing those carried by the regiment a brief glance at their early form and use will be made.

As a permanent force, the British army may be said to have existed from 1661. In 1664 it is stated (Sloane MS. 5247) that "A dragoner regiment of six troops, commanded by Colonel Worley, a Scotchman," exhibits all its six flags as of the guidon shape—that is, rounded ends with a slit in the centre. At this date a Captain Thomas Venn appears to have been the sole authority on colours and standards for the army.

In 1685 the four troops of Horse Guards each carried a standard and a guidon, of which at the Coronation of James II. the following is a description: "As to the standards: the cornet is of crimson damask doubled, being 2 feet 6 inches flying, and 2 feet 3 inches on the staff, etc. etc. The guidon is also of crimson damask, made up and embroidered in all respects as the cornet, from which it only differs in form, having a forked tail, from the point whereof to the staff is about 1 yard and 3 inches flying."

In 1661, when these four troops were formed, only one officer per troop, styled a cornet after the designation of his standard, was appointed; presumably, therefore, at this date only one standard per troop was carried. In 1674 however, a second cornet or guidon was appointed, and up to 1787 in each troop of Horse Guards two officers are shown as standard-bearers, one called major and cornet, the other major and guidon. *Question*—Was the officer called after the standard or the standard after the officer?

Turning to regiments of horse, it seems that about 1685 each troop carried *one* standard or guidon only. Those which at a later period became Dragoon Guards, such as the Earl of Peterborough's Regiment of Horse (now 2nd Dragoon Guards) and the Earl of Plymouth's Regiment of Horse (now 3rd Dragoon Guards), carried "standards": but those which at the above date may be called Light Cavalry regiments, such as the Royal Regiment of Dragoons commanded by the Lord Cornbury, the Princess Ann of Denmark's

Regiment of Dragoons commanded by John Berkely, Esq. (now the 4th Hussars), carried "guidons."

Up to the year 1743 the military authorities do not appear to have paid much attention to the pattern of, and devices on, standards and guidons. They were provided by the colonels of regiments, and were fashioned very much after the individual fancy of these officers. On the 14th September 1743 however, a Royal Warrant was issued regulating the colours and also the clothing of the army (copy of the warrant found in a work called 'Rudiments of War,' 2nd edition, 1782, published by Egerton), and the following instructions were given therein :—

No colonel is to put his arms, crest, device, or livery in any part of the appointments of his regiment.

The guidons of regiments of Dragoons to be of silk. The tassels and cords of the whole, to be of crimson silk and gold mixed.

The distinction of the third guidon, to be the figure 3 on a circular ground of red underneath the motto (where one is borne).

The size to be about 2 feet 10 inches in extreme length, 2 feet 4 inches deep.

The following detailed instructions are then given for the guidons of our regiment :—

The King's or first guidon to be crimson, with the rose and thistle conjoined, and crown over them, in the centre; his Majesty's motto, "Dieu et mon Droit," underneath. The white horse in a compartment in the first and fourth corners, and the rank of the regiment in silver characters on a ground of the same colour as the facing of the regiment (*i.e.* buff) in a compartment in the second and third corners.

The second and third guidons to be of the colour of the facing of the regiment (buff), with the badge of the regiment in the centre, or the rank of the regiment in silver Roman characters on a crimson ground within a wreath of roses and thistles on the same stalk, the motto of the regiment underneath. The White Horse on a red ground to be in the first and fourth compartments, and the rose and thistle conjoined upon a red ground in the second and third compartments.

On 1st July 1751, a further Royal Warrant was issued dealing with standards and guidons; it differed little from that of 1743, but a schedule was supplied with it giving full details as to colour, fringe, embroidery, badges, devices, and mottoes on the standards and guidons of each separate regiment. Those for the 9th Dragoons were as under :—

Colour of second and third guidon,—buff.

Embroidery on the three guidons,—silver.

Fringe ,, ,, —silver and blue.

Badge or device on the second and third guidons; rank of the regiment (IX. D.).

On the 19th December 1768, a further Royal Warrant dealing with standards and guidons was issued; little change was made, but the following dimensions were given. The lance of the guidons (except for Light Dragoons) to be 9 feet long, spear and ferrule included, the flag of the guidon to be 3 feet 5 inches to the end of the slit of the swallow-tail, and 2 feet 3 inches on the lance.

The following appears to have been about the cost at this time of a guidon and appointments: Guidon, £13, 13s. 6d.; pole, 11s. 4½d.; case, 10s.; guidon belts according to the richness of ornamentation. The regiment paid £3, 17s. 1d. for one.

Cavalry standards and guidons have never been treated with the same veneration as are the colours of the infantry, although as leading and rallying points, their objects seem identical. They were not consecrated, and when worn out they merely disappeared and new ones took their place without any peculiar notice.

They were not taken into the field during the Peninsular War, as in a letter from Colonel Clarke Kennedy, 1st Royal Dragoons, to Lieut. Siborne he says, "Please to recollect that none of the brigade took their standards into the field, they were left at home, by order" (Waterloo Letters); and on the Inspection Report of the 4th Dragoon Guards in 1812 it is stated, "Standards left on board the Admiral's ship at Lisbon."

From the fact that standards and guidons were not carried on the field of battle during the Peninsular War, their value and prestige seem to have lessened from this period in the eyes of the cavalry, and were further diminished by the following order; which deputed the carrying of the standards to troop-sergeant-majors:—

HORSE GUARDS, *November 30th*, 1822.

His Majesty has been pleased to command that the standards in cavalry regiments shall be carried in future by troop-sergeant-majors, and the Commander-in-Chief desires that the same may be observed accordingly.

H. TORRENS, *Adjt.-General.*

When regiments of Light Dragoons were formed into Hussars and Lancers, the use of them seems to have been abandoned, so that the regiment probably ceased to carry guidons in 1816.

In 1834 standards and guidons were entirely abolished, except for the Household Cavalry, seven regiments of Dragoon Guards, and three of Dragoons.

REGIMENTAL CALL AND MARCHES.

Regimental Call—

(Bugle)

(Trumpet)

Slow March—"The Men of Harlech."

Trot—"Encore."

Gallop—"The Irish Washerwoman."

Quick March—"The Soldiers' Chorus" (from *Faust*).

APPENDIX III.

ESTABLISHMENTS, 1715-1903.

Date.	No. of Troops.	Officers.	Non-commissioned Officers and Privates.						
1715	6	18	210						
Oct. 1715	6	18	270						
1778	6	22	222						
1797	8	31	669						
1798	9	36	..						
1801	10	38	..						
			Warrant Officers.	Sergeants.	Farriers.	Trumpeters.	Corporals.	Privates.	Horses.
Dec. 1802	8	31	..	28	..	8	32	608	436
May 1803	8	31	..	36	..	8	32	608	604
Feb. 1805	10	36	..	54	..	10	50	950	1064
Dec. 1809	10	41	..	34	..	10	30	570	..
1811 [1]	10	..	..	..	..	..	..	..	..
July 1814	8	36	..	..	8 added	..	..	..	453
Aug. 1821	6	31	..	..	..	..	..	282	253
1824 [2]	..	..	..	..	..	..	..	..	..
1831 [3]	..	..	..	..	..	..	..	..	..
April 1842	9	48	..	56	8	14	40	626	703 [4]
1853	8	..	← Rank and File—696 →						..
Dec. 1854 [5]	..	..	..	..	..	..	..	..	..
Mar. 1855 [6]	..	..	..	..	..	..	..	..	..
June 1857	10	..	..	..	..	..	..	..	..
Oct. 1859	8	34	..	40	8	9	32	537	428
June 1861	8	34	..	40	8	9	32	512	..
April 1863 [7]	..	..	..	..	..	..	..	..	..
Feb. 1871	..	28	..	42	8	9	32	315	384
Oct. 1874 [8]	7 1 depôt	30	..	38	6	8	27	461	456
Mar. 1885 [9]	7	24	2	33	6	6	24	384	396
Nov. 1885	8	24	2	41	8	8	32	378	300
May 1888	8	24	2	41	8	8	32	354	290
June 1889	8	..	..	..	..	..	..	..	281
Oct. 1889	8	24	2	40	8	8	32	382	300
	No. of Squadrons.								
Sept. 1890	4	24	of these 91 Ranks					591	424
April 1892	,,	24	,, 89 ,,					569	410
Aug. 1896	,,	21	2	43		7	32	391	360
	with depôt	2	..	7		1	5	93	
May 1897	3 1 Reserve Troop	23	2	475					360
April 1898	..	29	2	46		8	37	508	504
Oct. 1898	..	29	2	54		9	34	496	525
April 1903	4	29	2	49	4	9	36	497	527
							Officers' chargers		57

1 Between 1805 and this year the number of horses was reduced, and a sergeant-schoolmaster was added.
2 A hospital sergeant added.
3 An orderly-room clerk added.
4 These numbers include the depôt or recruiting troop.
5 A saddler sergeant added.
6 A riding-master, (a commissioned officer,) added.
7 A bandmaster sergeant added.
8 Depôt troop not included in above numbers.
9 Two troops depôt.

APPENDIX IV

LIST OF PLACES WHERE THE REGIMENT HAS BEEN QUARTERED AND ON "SERVICE," 1715-1903.

From		To		Place.
Year.	Month.	Year.	Month.	
1715	July	1717	June	Bedford, Warrington, Winchester.
1717	June	1718	May	Athlone (2), Portumna, Headford, Loughrea, Roscommon.
1718	May	1719	March	Collooney (2), Castlebar (1½), Ballinrobe (½), Boyle (1), Carrick ? (1).
1719	March	1719	Dec.	Castlereagh, Athlone, Roscommon.
1719	Dec.	1721		Castledawson (2), Newtown-Limavady (2), Dungannon (1), Lifford (1).
1721		1723		Killyleigh, Ballymena, Carrickfergus, Downpatrick, Loughrea, Castlebar.
1728		1729	June	Navan, Mullingar, Granard, Longford (2), Athlone.
1729	June	1729	June	Dublin.
1729	June	1730	June	Killough (3), Downpatrick, Saintfield, Carrickfergus.
1730	June	1731	May	Lifford, Castledawson (2), Newtown-Limavady (2), Dungannon.
1731	May	1732	June	Sligo (5), Collooney.
1732	June	1733	June	Castlebar, Headford, Portumna, Mount Talbot, Gort (2).
1733	June	1734	June	Longford (2), Athlone, Roscommon, Cavan, Belturbet.
1734	June	1734	Dec.	Philipstown (2), Granard (2), Mullingar, Navan.
1734	Dec.	1735	June	Dublin.
1735	June	1736	June	Saintfield, Carrickfergus, Downpatrick, Killyleigh, Killough (2).
1736	June	1737	June	Lifford, Dungannon, Castledawson (2), Hamilton's Bawn (2).
1737	June	1738	May	Sligo.
1738	May	1739	July	Castlebar (3), Gort, Loughrea, Headford.
1739	July	1740	June	Roscommon, Headford, Athlone, Portumna, Cavan, Belturbet.
1740	June	1741	June	Athlone (4), Limerick (2). To Dublin about May.
1741	June			From Dublin to Belturbet (2), Roscommon, Cavan, Athlone, Longford.
1742				No record.

Note.—The numerals in parentheses indicate the number of troops at each place.
Note.—The names of Irish "quarters" from 1717 to 1803 are written as now spelt in the "Postal Guide."

From		To		Place.
Year.	Month.	Year.	Month.	
1743				No record.
1744				Sligo, Collooney, Lifford, Ballyshannon.
1745				Lifford, Ballyshannon, Castlebar.
1746				Philipstown, Granard, Tullamore, Cappoquin, Mallow, Castlebar, Charleville.
1747				Dublin (3), Castlebar, Athlone, Granard.
1748				Athlone (3), Nenagh, Portumna, Granard.
1749				Athlone.
1750				No record.
1751		1753		Roscommon, Castlebar, Athlone, Granard.
1753		1754		Belturbet.
1754		1755	April	Sligo.
1755	April	1756		Castlebar (3), Portumna, Headford, Loughrea.
1756		1758		Clonmell (2), Ross, Tallow, Cappoquin.
1758		1760		Doneraile (1). No record of others.
1760		1761	June	Sligo.
1761	June	1763	March	Castlebar (3), Portumna, Headford, Loughrea.
1763	March	1764	March	Gort (3), Portumna, Loughrea, Nenagh.
1764	March	1765	Dec.	Belturbet, Cavan, Philipstown, Tullamore.
1765	Dec.	1766	June	Dublin.
1766	June	1767	Sept.	Castlebar (3), Athlone (2), Roscommon.
1767	Sept.	1768	April	Tipperary (2), Mallow (2), Doneraile (2).
1768	April	1769	April	Clonmell (3), Thurles (2), Fethard.
1769	April			Birr (3), Donaghmore, Nenagh.
1770				No record.
1771				No record.
1772	Jan.			Thurles.
1772	April	1773	April	Cappoquin (2), Tallow (2), Clogheen (2).
1773	April	1774	April	Clonmell (3), Clogheen, Roscrea, Ross.
1774	April			Kilkenny (2), Carrick-on-Suir (2), Callan, Ross.
1775				Carlow (3), Leighlinbridge, Tullow, Athy.
1776				Maryborough (3), Mountmallick (3).
1777	July	1778	June	Dublin.
1778	June			Carlow (4), Tullow (2).
1778	Oct.			In camp near Clonmell.
1779	Jan.			Tipperary (4), Doneraile (2).
1779	Aug.			Silgo (2). In camp, Ardfinnan (2).
1780				Athlone (4), Cloghan, Roscommon.
1781	July			Castlebar (2), Sligo (2), Ballinrobe, Dunmore.
1781	Oct.			Clonmell.
1782	Jan.			Charleville (5), Tralee.
1782	July			Charleville (4), Mallow (2).
1783	July			Ballinrobe (2), Castlebar (2), Loughrea (2).
1783	Oct.			2 troops Loughrea to Galway.
1784	Jan.			2 troops Galway returned to Loughrea.
1784	July			Dublin.
1784	Oct.			Carlow (4), Athy, Ross.
1785				Belturbet (4), Cavan, Augher.
1786	July			Maryborough (3), Mountmallick (2), Thurles.
1786	Oct.	1787	July	Clonmel (2), Cappoquin (2), Clogheen (2).
1787	July	1788	July	Athlone (3), Portumna, Roscommon, Loughrea.
1788	July	1790	March	Charleville (2), Nenagh (2), Mallow (2).
1790	March	1791	Jan.	Carlow, Athy, Tullow.
1791	Jan.	1792	Jan.	Headquarters, Carlow.
1792	Jan.			Longford (2), Sligo.

From		To		Place.
Year.	Month.	Year.	Month.	
1793				No record.
1794	Feb.			Headquarters, Dublin.
1794	July	1796	Jan.	,, Longford.
1796	Jan.	1796	Sept.	,, Tullamore.
1796	Sept.	1798	Jan.	,, Dublin.
1798	Jan.			,, Carlow (9).
1799	May			,, Kilkenny.
1799	Sept.			,, Maryborough.
1800	Nov.	1801	Dec.	,, Ballinrobe.
1801	Dec.	1802	Dec.	Ballinrobe, Limerick, Mallow.
1803	May	1804	May	Ipswich.
1804	May	1805	June	Hounslow.
1805	June	1806	March	Guildford, Blatchington, and other places in Sussex.
1806	March	1806	Oct.	Hythe, Romney, Ramsgate, Sandwich.
1806	Oct.	1807	Nov.	Monte Video, Buenos Ayres, &c., S. America. 2 troops left at Coventry.
1807	Nov.	1809	July	Guildford, Hounslow.
1809	July	1809	Oct.	The Island of Walcheren.
1809	Oct.	1811	March	Exeter.
1811	July	1813	May	The Peninsular War, Portugal. Depôt, Coventry.
1813	May	1814	March	Northampton and Coventry.
1814	March	1814	Sept.	Exeter.
1814	Sept.	1815	May	Dublin.
1815	May	1816	March	Manchester and Nottingham.
1816	March	1817	June	Hounslow and Kensington.
1817	June	1818	March	Canterbury.
1818	March	1819	June	Brighton.
1819	June	1820	June	Nottingham.
1820	June	1821	June	Romford.
1821	June	1822	March	Canterbury.
1822	March	1823	March	Brighton and Radipole Barracks.
1823	March	1824	June	Cahir.
1824	June	1825	March	Dublin.
1825	March	1826	March	Ballinrobe.
1826	March	1827	March	Glasgow.
1827	March	1830	April	York.
1830	April	1832	June	Hounslow.
1832	June	1833	April	Dublin.
1833	April	1834	March	Longford.
1834	March	1835	May	Newbridge.
1835	May	1836	May	Coventry.
1836	May	1837	April	Edinburgh.
1837	April	1838		Glasgow.
1838				Leeds and Nottingham.
1839		1840	April	Ipswich.
1840	April	1841	April	Hounslow.
1841	April	1842	May	Dorchester.
1842	May	1843	Jan.	Fort William, East Indies.
1843	Jan.	1845	Oct.	Cawnpore.
1843	Nov.	1844	Jan.	A portion of the Regiment with the Army of Gwalior.
1845	Oct.	1845	Dec.	Meerut.
1845	Dec.	1846	March	Sutlej Campaign, 1st Sikh War.
1846	March	1848	Sept.	Meerut.
1848	Sept.	1849	Feb.	With the Army of the Punjab, 2nd Sikh War.

From		To		Place.
Year.	Month.	Year.	Month.	
1849	Feb.	1851	Feb.	Wuzeerabad.
1851	Feb.	1857	May	Umballa.
1857	May	1858	June	Joined the force for the capture of Delhi. In the field during the Indian Mutiny.
1858	June	1858	Sept.	Umballa.
1858	Sept.	1859	May	In the field to the end of the Mutiny.
1859	May	1860	Sept.	Exeter.
1860	Sept.	1862	Aug.	Aldershot.
1862	Aug.	1864	March	Brighton (6), Shorncliffe (2).
1864	March	1864	Sept.	Newbridge.
1864	Sept.	1865	April	Dublin.
1865	April	1866	July	Dundalk (6), Belfast (1), Belturbet (1).
1866	July	1866	Oct.	The Curragh.
1866	Oct.	1867	Oct.	Dublin.
1867	Oct.	1868	Oct.	Cahir, Cork, Waterford, Clonmell.
1868	Oct.	1869	July	Newbridge.
1869	July	1870		Hounslow, Hampton Court, Kensington.
1870		1872	May	Aldershot.
1872	May	1872	Sept.	Woolwich.
1872	Sept.	1874	July	York.
1874	July	1875	Jan.	Colchester.
1875	Jan.	1878	Oct.	Sialkote, India.
1878	Oct.	1879	June	With the Peshawar Valley Field Force.
1879	June	1879	Sept.	Sialkote.
1879	Sept.	1880	Dec.	The Afghan Campaign.
1880	Dec.	1885	March	Umballa.
1885	March	1885	April	Rawal-Pindi. Durbar for Abdul Rahman, the Ameer of Afghanistan.
1885	April	1885	Oct.	Umballa.
1885	Oct.	1886	Aug.	Shorncliffe.
1886	Aug.	1888	April	York.
1888	April	1890	Aug.	Manchester, Seaforth (Liverpool).
1890	Aug.	1892	June	Dundalk, The Curragh, Belfast.
1892	June	1894	June	Newbridge, The Curragh.
1894	June	1896	Aug.	Aldershot.
1896	Aug.	1897	May	Pietermaritzburg. Depôt, Canterbury.
1897	May	1898	March	Ladysmith. Reserve Troop, Pietermaritzburg.
1898	March	1899	Sept.	Muttra.
1899	Oct.	1902	March	South African War. Reserve Squadron, Muttra.
1902	April	1902	Oct.	Sialkote.
1902	Oct.	1903	Jan.	Delhi. Coronation Durbar, King Edward VII.
1903	Jan.			Rawal-Pindi.

APPENDIX V.

ALTERATIONS IN DRESS, HORSE FURNITURE, &c. 1715-1903.

1751. By order of King George II., dated July 1751, the following was the order of dress, &c., for the regiment:—

Coats.—Scarlet, double-breasted without lapels, lined with buff; slit sleeves, turned up with buff; the button-holes ornamented with narrow white lace; the buttons of white metal, set on two and two; a long slash pocket in each skirt, and a white worsted aiguillette on the right shoulder.

Waistcoat and Breeches.—Buff colour.

Hats.—Three-cornered cocked-hats, bound with silver lace, and ornamented with a white metal loop and a black cockade.

Boots.—Of jacked leather, reaching to the knee.

Cloaks.—Of scarlet cloth, with a buff cape, and lined with buff shalloon; the buttons set on two and two, upon white frogs or loops, with a blue stripe down the centre.

Horse Furniture.—Of buff cloth; the holster-caps and housings having a border of white lace with a blue stripe down the centre; IX. D, embroidered upon the housings on a red ground, within a wreath of roses and thistles; $\frac{\text{II.}}{\text{G.R.}}$ embroidered on the holster-caps with a crown over it, and IX. D underneath.

Officers.—Distinguished by silver lace; their coats and waistcoats bound with silver embroidery, the button-holes worked with silver, and a crimson silk sash worn across the left shoulder.

Quartermasters.—To wear a crimson sash round the waist.

Sergeants.—To have narrow silver lace on the cuffs, pockets, and shoulder-straps; silver aiguillettes, and buff and blue worsted sashes tied round the waist.

Drummers and Hautbois.—Clothed in buff, lined and turned up with red, and ornamented with white lace, having a blue stripe down the centre; red waistcoats and breeches.

1758. Clothing is described as red, not scarlet: faced buff. Horse furniture buff.

In consequence of being made a Light Dragoon regiment, a change in uniform and equipment was ordered from this year; but owing to being quartered in Ireland, it was not finally effected till 1804. 1783.

The clothing of Light Dragoon regiments was changed from scarlet to blue in this year. 1784.

On return to England in this year, it was fully clothed and equipped, as a Light Dragoon regiment. 1804.

The facings on the clothing were changed (from buff?) to crimson, and the pattern of the clothing and appointments was also changed. 1811.

In December, pantaloons and hussar boots were discontinued, grey overalls and short boots taking their place. 1815.

The detachments of the 9th, 12th, 16th, and 23rd Light Dragoons, which assembled in London in December of this year to be instructed in the "Exercise of Lancers" under Captain Peters of the regiment, were ordered to be clothed as follows:— 1816.

Jacket.—As for a light dragoon of 1812.

Trousers.—Cossack pattern; very full in the thigh, tightening towards the knee, and fitting close over a short boot.

Lance Cap.—Very high, with a square top, made of cane, and covered with cloth, the same colour as the facings of each regiment (9th L.D. crimson). A brass plate in front, and a plume on the top (plume probably red and white).

Arms.—From this year the carbine was no longer carried by the dragoon.

The following are the dress regulations for the regiment, dated Adjutant-General's Office, Horse Guards, 25th April 1822:— 1822.

Full Dress—

" *Jacket.*—Blue cloth, lapels, collar, and cuffs, the colour of the facings of the regiment (crimson).

" *Trousers.*—Sky blue. Two stripes of lace $\frac{3}{4}$ inch wide each.

" *Spurs.*—Yellow or white metal $2\frac{1}{4}$ inches in the neck, including the rowels.

" *Chaco.*—Black beaver, bell shape; 8 inches deep; plume red and white, drooping about 23 inches. This however was not worn, but the lance cap as described for 1816.

" *Girdle.*—Gold lace; 3 inches wide, with two $\frac{3}{8}$-inch crimson stripes. To fasten externally with three gold cord loops, and embroidered olivets.

" *Sabre.*—Steel.

" *Waistbelt.*—Gold lace, $1\frac{1}{4}$ inches wide.

" *Sabretache.*—Purple leather inside, blue cloth outside.

" *Pouch Belt.*—Gold lace, 2 inches wide, with a $\frac{1}{2}$-inch silk

1822. *Full Dress, continued—*
stripe in centre; silver engraved plates, with chains and pickers.
" *Pouch Box.*—Leather of, to be black.

Undress—
" *Pouch Belt.*—Plain white buffalo leather 2½ inches wide.
" *Pouch Box.*—Plain black patent leather: flap 6½ inches wide, 4½ inches deep.
" *Forage Cap.*—Blue cloth with welts and plaits; band 1½ inch deep; button at top; black patent leather peak.

1829. The colour of the overalls was altered from blue-grey to dark-grey; lapels on the jacket were abolished, and a black plume was substituted for a red and white one.

1831. By warrant of 18th May the clothing was changed as under:—

Coat.—Scarlet, double breasted; collar, Prussian blue.
Epaulettes.—Blue cloth strap, richly embroidered with gold.
Cap.—Blue cloth; trencher top 10 inches square; plume, black cock's tail, drooping 16 inches in front, 9 inches at the back.
Trousers.—Blue cloth, with two stripes of gold lace.
Spurs.—Of yellow metal two inches long, with dumb rowels.
Girdle.—Gold lace 2½ inches wide, with two stripes of blue silk in it. To fasten with hooks and eyes.
Pouch Belt.—Gold lace, 2 inches wide, with a blue silk stripe in centre ½ inch wide; silver plates with chains and pickers.
Pouch Box.—The leather of scarlet; a gold embroidered edging round the top; a solid silver flap 7½ inches wide, 2¾ inches deep, with a solid raised W.R. and crown, gilt. (These directions, I do not think, were strictly carried out. The present cypher monogram of A.R. reversed, would, I think, be used from 1830.)

Undress—
" *Trousers.*—Blue cloth, two scarlet cloth stripes.
" *Spurs.*—Steel, 2½ inches long, sharp rowels.
" *Stable Jacket.*—scarlet, shoulder straps, gold cord.
" *Forage Cap.*—Blue, with a gold band.
" *Cloak.*—Blue, scarlet collar and lining.
" *Gloves.*—White.

1833. It was ordered that a sabretache should be worn on dismounted duties.

1837. It is recorded in the official records of the regiment, by Horse Guards letter of the 12th August, that it was to retain on its buttons and appointments the cypher W.R. instead of adopting that of V.R. as generally ordered on the 31st July 1837. I think this has been a clerical error in compiling the records; and it was intended that the cypher monogram of A.R. reversed (which by a casual observer might easily be mistaken for that of the letters W.R.) was that to be retained.

The colour of the clothing was changed from red to blue, and the facings to scarlet. 1840.

By a clothing warrant this year the following was the dress ordered :— 1846.

Jacket.—Blue, double breasted.

Cap.—According to regimental pattern. (The same as in 1831.) Plume, black; drooping cock's-tail feathers 14 inches long and 4 inches on the mount. In India, black horse-hair of the same length to be worn.

Girdle.—Gold lace 2½ inches wide, with two crimson silk stripes in it.

Sabretache.—Scarlet cloth face, edged with gold lace. At the bottom an embroidered scroll with the regimental badge in velvet above it; above this a gold embroidered double cypher monogram of A.R. reversed.

With the exception of the entry quoted in the year 1837, this is the first official recognition I can trace of authority for our present cypher monogram; but I think it is almost certain that when the regiment in July 1830 was authorised to assume the title of "Queen's Royal," it would from that date bear the monogram of the Queen after whom it was named—"Adelaide Regina."

Pouch Box.—Leather of, to be blue; silver flap with double cypher monogram of A.R.

Pouch Belt.—Is not mentioned in the warrant, but I think, as the colour of the silk in the girdle was changed from blue to crimson that in the belt would also correspond.

Shabracque.—Blue cloth.

Gauntlets.—White leather.

Undress—

" *Trousers.*—Blue cloth. Two scarlet stripes.

" *Stable Jacket.*—Blue cloth.

" *Waist Belt.*—Brown morocco leather.

" *Forage Cap.*—Blue cloth, oak-leaf band 1¾ inch wide, gold embroidered peak, oilskin cover.

" *Frock Coat.*—Blue cloth.

" *Shabracque.*—Sheepskin or black lambskin.

" *Gloves.*—White leather.

By Horse Guards Order of the 27th November, the stripes on the overalls were changed from scarlet cloth to yellow lace. This presumably refers to the "undress" trousers. 1849.

By a clothing warrant of this year, the following alterations were made :— 1855.

Jacket.—Blue, with blue lapels, and the following badges of rank were instituted :—

A Colonel.—A crown and star.

A Lieut.-Colonel.—A crown.

M

1855. *A Major.*—A star.

A Captain.—A crown and star.

A Lieutenant.—A crown.

A Cornet.—A star.

The badges to be worn at each end of the collar, and the collars and cuffs of field officers to have 1 inch lace round them.

1856. The Lancer cap now in use was made a "sealed" pattern this year.

1857. By a clothing warrant of this year the following changes were made :—

Tunic.—Blue, double breasted, scarlet lapels.

Boots.—Wellington.

Plume.—Black and white cock's-tail feathers, 12 inches long standing $4\frac{1}{2}$ inches above the top of the cap.

Undress—

„ *Trousers.*—Blue cloth; two pale yellow stripes.

„ *Forage Cap.*—A gold band of regimental lace $1\frac{3}{4}$ inches wide; gold embroidered peak and an oilskin cover.

Saddlery—Saddle.—Hussar pattern; brown hogskin.

„ *Collar Chain.*—Of steel.

„ *A Crupper.*

„ *Valise.*—Blue cloth, with Arabic number and letter on each end 9. *L*.

„ *Shabracque.*—To have on each of the fore corners V.R., and a crown in gold. On each of the hind quarters, a crown over V.R. within a garter, with the words "Ninth Lancers" round it; and two lances crossed under the garter, all in gold.

1861. *Pouch Box and Belt.*—The flap and ornaments of the pouch belt to be of gold, *i.e.* silver-gilt. This change had however, been made regimentally about 1857.

1864. The coloured collar of the cloak was abolished.

1874. Pantaloons and knee boots with steel spurs, were adopted for mounted duties, in place of Wellington boots and box spurs. The crimson silk stripes, in the lace for the pouch belt, and sword slings were abolished. This change appears to have been made regimentally about 1859. It was decided in June of this year to retain the blue frock coat as undress, in place of a patrol jacket.

1875. The present plume for the lance cap was "sealed" as a pattern on the 29th March.

1878. The leather of the pouch was altered to red.

In a clothing warrant of this year it was ordered that the mess waistcoat should be of a regimental pattern.

In the dress regulations of this year the ornamentation on the shabracque is described in more correct Heraldic terms, as follows:—On the fore corners, the Royal Cypher V.R. and

crown, in gold. On the hind corners, surmounted by the 1878.
Imperial crown, the Royal Cypher within a garter, bearing the words "Ninth Lancers" on crossed lances.

The regiment was armed with the Martini-Henry carbine in place of the pistol.

In May of this year the following alterations were officially 1883.
made :—

Tunic.—Blue, with plastron front, which is reversible, showing scarlet for review order and levée dress, blue for marching order.

The origin of this plastron front is of interest, dating back to the middle of the twelfth century (1170). It was then called the "plastron de fer," and was a steel plate thus shaped,

which was worn by knights in armour, between the hauberk or haubergen and the gambeson; the hauberk being a complete suit of mail, and the haubergen a coat of mail only. The gambeson was a tunic made of buckskin, filled with cotton and quilted, in order to take the pressure of the mail armour off the chest. It was sometimes worn beneath the gambeson.

It is related that in a combat between Richard Cœur de Lion and a knight called William de Barras, they charged each other so fiercely that their lances pierced through their shields, hauberks, and gambesons; and were only prevented by the "plastrons" from entering their bodies.

In later years this plastron was called a "gorget."

Plume.—A black-and-white horse-hair plume was brought into use (presumably for officers only) 12 inches long $4\frac{1}{2}$ inches above the top of the lance cap.

The length of the black-and-white feather plume is ordered to be 14 inches in front and 7 inches behind.

A field-service forage cap, made of blue cloth, to fold flat, was this year authorised for use.

The use of gold lace on the pantaloons of officers was dis- 1894.
continued this year, yellow cloth stripes (2) taking their place.

In February, gold lace on officers' overalls was abolished, 1902.
yellow cloth stripes being substituted.

The old mess jacket and mess waistcoat, hooking up to the neck were abolished, being replaced by a jacket with a roll collar, and an open waistcoat.

APPENDIX VI.

COLONELS OF THE REGIMENT.

Major-General Owen Wynne . . .	22nd July 1715.
Major-General James Crofts . . .	6th July 1719.
Major-General the Viscount Molesworth .	29th May 1732.
Major-General Sir John Cope . . .	27th June 1737.
Colonel John Brown	10th May 1742.
Colonel Henry de Grangue . . .	1st April 1743.
Lieut.-General George Reade . . .	1st November 1749.
Colonel John Jorden	2nd April 1756.
Major-General Philip Honywood . .	22nd May 1756.
Lieut.-General Henry Whitley . . .	6th April 1759.
Major-General James Johnston . . .	17th January 1771.
General Flower Mocher	27th April 1775.
General Sir James St. Clair Erskine, Bart. .	1st August 1801.
(who from January 1805 became the Earl of Rosslyn, G.C.B.)	
Lieut.-General Samuel Need . . .	30th January 1837.
Lieut.-General Sir James Wallace Sleigh, K.C.B.	24th August 1839.
General Sir James Hope Grant, G.C.B. . .	6th February 1865.
General Sir Archibald Little, G.C.B. . .	8th March 1875.
Lieut.-General Sir William Drysdale, K.C.B. .	11th June 1891.
Major-General Henry Augustus Bushman, C.B. .	8th August 1900.

MAJOR-GENERAL OWEN WYNNE.

Appointed Colonel 22nd July 1715.

He was third son of Owen Wynne, who, leaving Wales, settled in Ireland about 1688.

He obtained a commission in the army of James II. in the spring of 1688, but being of the Protestant religion; when the Revolution broke out he took service under the Prince of Orange, and proceeding to Ireland with one of the regiments sent to the relief of Londonderry, was detached to help the Inniskilling men in their resistance to James II.

When the Inniskilling bands were formed into regiments under King William, he obtained a commission in Colonel James Wynne's Dragoons (5th Royal Irish), with which regiment he saw much service in Ireland and in Flanders, and was appointed Lieut.-Colonel, 20th July 1695. He served in the Netherlands under the Duke of Marlborough, was promoted Colonel in 1703, and in 1705 was commissioned to raise a regiment of foot, of which he was appointed Colonel. This regiment served in the campaigns of 1710 and 1711 under the Duke of Marlborough, and that of 1712 under the Duke of Ormonde, but at the Peace of Utrecht (1713) it was disbanded. He was promoted to the rank of Brigadier-General in 1706, and on 1st January 1709 to that of Major-General.

When the army was augmented in 1715, owing to the Jacobite revolt, Major-General Wynne was commissioned to raise a regiment of Dragoons—the 9th (Queen's Royal) Lancers—being appointed Colonel of it.

In 1719 he was moved to the 5th Horse (4th Dragoon Guards). On the 10th March 1726 he was promoted to the rank of Lieut.-General, and was also appointed one of His Majesty's Privy Councillors.

He represented the borough of Ballyshannon in Parliament from 1715 to 1727, and from that year to his death in February 1737, was M.P. for the county of Sligo.

In 1728 he was Commander-in-Chief of His Majesty's forces in Ireland, and in August 1732 was transferred from the colonelcy of the 5th Horse to that of the 5th Royal Irish Dragoons—in which regiment he had previously served—retaining this appointment till February 1737.

In 1736 he was appointed Governor of Londonderry and Culmore Forts.

Francis Cocksedge, agent to his regiment, left General Wynne a nice estate near Blackrock, Dublin, which his nephew Owen sold.

General Wynne several times refused a peerage, thinking it more honourable to be one of the first commoners than the last-made peer.

He was one of the many caught in the South Sea Bubble (1720), losing £10,000 in the venture.

Colonel James Crofts.

Appointed Colonel 6th July 1719.

This officer was a natural son of the celebrated James, Duke of Monmouth, who raised the standard of revolt in 1685 and was beheaded in July of that year.

He entered the army in the early part of Queen Anne's reign (1702-3), became a Colonel in 1706, and in 1718 succeeded Sir Robert Rich, in command of a regiment of Dragoons, which was disbanded that year.

In 1719 he was appointed Colonel of the 9th Lancers, and in 1727 was promoted a Major-General. He died at his house in Downing Street, London, in March 1732.

Richard, Viscount Molesworth.

Appointed Colonel 29th May 1732.

He was the third son of Robert Molesworth, Ambassador at the Court of Denmark for many years, who was advanced to the Peerage of Ireland 16th July 1716, as Baron of Philipstown and Viscount Molesworth. He was intended for the law, but on the breaking out of the War of the Spanish Succession, he proceeded to Holland, and served for a few weeks as a volunteer with Major-General the Earl of Orkney's Regiment of Foot (First Royal Foot), until his lordship, who was an intimate friend of his father's, procured him a commission. He served in the Netherlands and Germany, showing great personal bravery, talent, and strict attention to duty, for which he was rewarded with a commission as Captain of Horse, and was appointed Aide-de-Camp to the Duke of Marlborough.

During the battle of Ramillies on 23rd May 1706, while leading a body of cavalry to the charge, the Duke was cannoned against while in the act of jumping a ditch, and was thrown from his horse. Captain Molesworth was the only officer near; he at once dismounted, gave his horse to the Duke, helped him to mount, and was then left to face the French cavalry (dismounted). They luckily however, were so intent upon pursuing the Duke, that Captain Molesworth escaped with a few sabre cuts on the shoulders and arms. He recovered the Duke's horse and joined his suite. Captain Molesworth's horse proving rather restive, the Duke got back on to his own; while so doing a cannon ball took off the head of Colonel Bringfield, his equerry, who was holding the stirrup.

He continued to serve with distinction under the Duke of Marlborough, and was transferred from the cavalry to His Grace's regiment of Foot Guards (1st).

In May 1707 he was appointed Captain and Lieut.-Colonel in the 2nd Foot Guards, and in 1710 was promoted to the Colonelcy of a regiment of foot in succession to Colonel Moore. While serving in Flanders he was blown up by a mine, but not seriously injured.

In 1711 his regiment was ordered to Spain, but on the Peace of Utrecht (1713) it was disbanded.

In 1714 he was appointed Lieut.-General of the Ordnance in Ireland.

In the summer of 1715 he was one of the officers called upon to raise a regiment of Dragoons, at the head of which he served during the rebellion of the Earl of Mar. He was employed under Major-General (afterwards Lord) Carpenter, in pursuit of the division of the rebel army under General Foster and Brigadier-General Mackintosh in their retreat from Preston, being wounded on this occasion.

His regiment of Dragoons, together with ten others, was disbanded 1717-1718.

In 1725 he was appointed Colonel of the 27th Regiment of Foot. On the death of his brother, 18th February 1726, he succeeded to the title of Viscount Molesworth of Swords, County Dublin, and Baron of Philipstown, King's County. In 1732 he was appointed Colonel of the 9th Lancers. In 1733 he was sworn a member of the Privy Council. In 1735 he was promoted Major-General, and was transferred in 1737 to the Colonelcy of the 5th Royal Irish Dragoons.

In 1739 he was promoted Lieut.-General, became Master-General of the Ordnance in Ireland in 1741, was promoted General in 1746, was appointed Commander-in-Chief in Ireland in 1751, and became a Field Marshal in 1757. He was married twice, his first wife being Jane Lucas (died 1742); his second, whom he married 1743, being Mary, daughter of the Reverend William Usher, Archdeacon of Clonfert, by whom he left a son. After a lengthy, varied, and brilliant career he died 12th October 1758.

Brigadier-General Sir John Cope.

Appointed Colonel 27th June 1737.

He was appointed a Cornet of Dragoons, March 1707, and served under the Duke of Marlborough. In 1715 he was promoted to be a Captain and Lieut.-Colonel in the 2nd Foot Guards, in April 1720 to the Lieut.-Colonelcy of the second troop of Horse Grenadier Guards, and in 1721 was promoted to the rank of Colonel.

On the 10th November 1730, King George II. rewarded him by promotion to the Colonelcy of the 39th Regiment of Foot, from which in 1732 he was moved to the 50th Regiment of Foot, and in 1735 was promoted to the rank of Brigadier-General.

He served on the staff in Ireland for several years. In 1737 he was appointed Colonel of the 9th Lancers, being transferred in 1741 to the Colonelcy of the 7th Dragoons, having meanwhile (1739) been promoted a Major-General.

In 1742, he proceeded to Flanders with the army under the Earl of Stair, and in 1743 was at the head of the second line of cavalry at the Battle of Dettingen; for his distinguished conduct on this occasion he was made a Companion of the Order of the Bath. In the same year he was promoted to the rank of Lieut.-General.

In 1745, he was Commander-in-Chief in Scotland, and 2000 troops under his command were defeated at the battle of Prestonpans on 21st September, by the Young Pretender. He died in 1760.

Colonel John Brown.

Appointed Colonel 10th May 1742.

He obtained his first commission as Cornet in a regiment of horse 5th August 1704, and served in several campaigns under the Duke of Marlborough. He was for several years Lieut.-Colonel of the 4th Dragoons and afterwards of the King's Horse (1st Dragoon Guards). On the 10th May 1742 he was rewarded with the Colonelcy of the 9th Dragoons, from which in April 1743 he was moved to the 5th Horse (4th Dragoon Guards). He became Major-General 1754, Lieut.-General 1758, and died in 1762.

Colonel Henry de Grangue.

Appointed Colonel 1st April 1743.

He entered the army 1695, served in the wars of Queen Anne, and became Lieut.-Colonel of the Royal Regiment of Dragoons.

When the dispute concerning the succession to the throne of Hungary and Bohemia involved war in Europe, he was on the 21st

January 1741 appointed Colonel of a newly raised regiment of foot, numbered the 60th. In October 1742 he was moved to the 30th Foot.

On the 1st April 1743 he was appointed Colonel of the 9th Dragoons, being moved in November 1749 to that of the 4th Irish Horse (7th Dragoon Guards), an appointment he held till his death in 1754.

LIEUT.-GENERAL GEORGE READE.

Appointed Colonel 1st November 1749.

He obtained his first commission August 1703, and served under the Duke of Marlborough.

In June 1733 he was promoted from Major in the 1st Regiment of Foot Guards to the Colonelcy of the 29th Foot, and in 1739 was transferred to the 9th Foot.

He was promoted a Major-General 1743 and a Lieut.-General in 1747. In November 1749 he was given the Colonelcy of the 9th Dragoons, which appointment he retained till his death in 1756.

COLONEL JOHN JORDEN.

Appointed Colonel 2nd April 1756.

He obtained his first commission in December 1704, and served in the 10th Dragoons from July 1715, being appointed Lieut.-Colonel Commanding in February 1741. He was appointed Colonel of a regiment of marines March 1747, and in 1748 to that of the 15th Foot. In April 1756 he was appointed Colonel of the 9th Dragoons, but died the following month.

Major-General Philip Honywood

Appointed Colonel 22nd May 1756.

Was a descendant of Robert Honywood of Charing, County Kent, and Marks Hall, Essex; to the latter of which places he succeeded in 1758.

He entered the army in 1735, and in 1741 was appointed a Major in the 3rd Dragoons.

In 1742 he proceeded with his regiment to Flanders, and showed signal gallantry at the battle of Dettingen in 1743, where it engaged the French household troops and suffered severely. Major Honywood received five wounds, and was left for dead on the field of battle for six hours.

He recovered of his wounds, and was present at the battle of Fontenoy in 1745.

The rebellion in Scotland breaking out in the autumn of this year, he returned with his regiment to England, and was employed in the pursuit of the Highlanders. He overtook their rear-guard on Clifton Moor, when a sharp action took place and he was wounded in the shoulder.

In March 1752 he was promoted a Colonel in the army, and in 1755 was rewarded with the Colonelcy of the 20th Foot. In May 1756 he was promoted to that of the 9th Dragoons, and in 1759 was transferred to the 4th Horse (7th Dragoon Guards).

During the Seven Years' War he commanded a brigade of cavalry under Prince Ferdinand of Brunswick, and particularly distinguished himself at the battle of Warbourg on the 31st July 1760.

He was promoted a Major-General in May 1758, a Lieut.-General December 1760, and subsequently a General.

In 1782 he was appointed Colonel of the 3rd Dragoon Guards.

He was for many years Governor of Hull, and Member of Parliament for Appleby (Lincolnshire).

He died 29th March 1785.

Lieut.-General Henry Whitley.

Appointed Colonel 6th April 1759.

He served in Lieut.-General Bland's regiment of dragoons, and was appointed as Lieut.-Colonel of the 10th Dragoons in March 1747, vice Jorden.

He was rewarded with the Colonelcy of the 9th Dragoons, 6th April 1759.

He became a Major-General 1761, and a Lieut.- General 1770.

He died January 1771.

Major-General James Johnston.

Appointed Colonel 17th January 1771.

He was appointed Cornet 13th Dragoons 5th October 1736, and was transferred to the Royal Dragoons in 1739, in which regiment he rose to the rank of Major.

He was promoted to the Lieut.-Colonelcy of the 13th Dragoons 2nd April 1754. In 1759 he returned to the Royal Dragoons, and proceeding to Germany with that regiment was present at many battles and skirmishes under Prince Ferdinand of Brunswick.

He distinguished himself at the battle of Warbourg 1760, and was wounded at that of Campen.

He commanded a brigade of cavalry in Germany in 1762. In 1763 he was appointed Governor of the island of Minorca, and in 1770 was made a Major-General.

In January 1771 he was rewarded with the Colonelcy of the 9th Dragoons; in 1774 was appointed Governor of Quebec, and in 1775 was transferred to the Colonelcy of the 1st Irish Horse (4th Dragoon Guards) in succession to his cousin, General James Johnston.

He was reported to be the best swordsman of his time.

He became a General in 1793, and died December 1797 at Hampton, from whence he was moved with great state and buried in Westminster Abbey on the 21st December 1797.

General Flower Mocher

Appointed Colonel 27th April 1775.

Was for many years an officer in the second troop of horse, Grenadier Guards, in which regiment he became a Major in 1754.

In 1759 he was promoted to the Lieut.-Colonelcy of the 3rd Dragoon Guards, with which regiment he served in Germany during the Seven Years' War.

In 1775 (27th April) King George III. rewarded him with the Colonelcy of the 9th Dragoons, which he retained till his death in 1801, having been promoted Major-General 1777, Lieut.-General 1782, General 1796.

General Sir James St. Clair Erskine, Bart.,

From January 7, 1805, *the Earl of Rosslyn.*
Appointed Colonel 1*st August* 1801.

This officer was the eldest son of Lieut.-General Sir Henry Erskine, Bart., of Alva; and was born in 1762. His early services were so diversified that a tabulated statement will best explain them.

Cornet, 1st troop (now 1st Regiment) of Life Guards	12th March 1778.
Lieutenant, 21st Light Dragoons . .	December 1779.
Captain, 19th " . .	1780.
" 14th " . .	1781.
A.D.C. to Lord-Lieutenant of Ireland, and afterwards Assistant-Adjutant-General in Ireland.	
Major, 8th Light Dragoons . . .	1783.
Lieut.-Colonel, 12th Light Dragoons . .	1792.

In 1789 he assumed the additional surname of St. Clair.

In 1793 he served with the 12th Light Dragoons at Toulon, was Assistant-Adjutant-General to the Forces in the Mediterranean, and was present at the capture of the island of Corsica 1794.

In 1795 he was promoted Colonel in the army, and appointed an A.D.C. to King George III.

In 1796 he was appointed a Brigadier-General in Portugal. In January 1798 he was made a Major-General, and proceeding from Portugal to the Mediterranean was present at the reduction of the island of Minorca, where he was Commander-in-Chief until the end of 1799, when returning to England, he was rewarded with the Colonelcy of the Sussex Fencible Cavalry.

On the 1st August 1801, he was appointed Colonel of the 9th Light Dragoons.

On the death of his maternal uncle he succeeded to the title of the Earl of Rosslyn in the county of Midlothian, and Baron Loughborough of Loughborough in the county of Leicester, as the 2nd Earl.

In January 1805 he became a Lieut.-General, and was employed on the Staff in Ireland, North Britain, and Portugal.

In 1807 he served in the expedition to Denmark under Lord Cathcart.

In 1809 he commanded a division of light infantry in the Walcheren expedition, and on his return therefrom was again employed on the Staff in North Britain.

In 1812 he was appointed to command the troops in the Kent district, and continued to hold this appointment until June 1814, when he was promoted General.

He had a large and varied experience in military matters, was constantly consulted by the authorities, more especially on those connected with the cavalry service.

He was for twenty years a member of the House of Commons before he succeeded to the peerage.

In 1820 he was appointed a Knight Grand Cross of the Order of the Bath (G.C.B.); in 1829 was sworn a Privy Councillor and appointed Keeper of the Privy Seal. In 1834 he was appointed Lord President of the Council during Sir Robert Peel's administration.

He was Councillor of State to the King in Scotland, and Lord-Lieutenant of Fifeshire.

He died 18th January 1837, at Dysart House, Fifeshire; having been Colonel of the 9th Lancers for thirty-six years.

His eldest son, James Alexander, Lord Loughborough; who was born in February 1802, joined the regiment as Cornet 25th February 1819; and commanded it from March 1828 to January 1842.

Lieut.-General Samuel Need.

Appointed Colonel 30th January 1837.

He was appointed Cornet in the 1st Dragoon Guards 1784, Lieutenant in 1786, and Captain in 1793.

In 1796, he was promoted to the Majority of the 27th (afterwards 24th) Light Dragoons.

While on the voyage to the East Indies he was present at the capture of the armament which the Dutch had equipped to retake the Cape of Good Hope.

He served in India in all Lord Lake's campaigns, during the last two of which he commanded a brigade of cavalry.

He became a Colonel in 1802, Major-General in 1814, and Lieut.-General in 1830. In January 1837 King William IV. conferred on him the Colonelcy of the 9th (Queen's Royal) Lancers, which he retained till his death in 1839.

General Sir James Wallace Sleigh, K.C.B.

Appointed Colonel 24th August 1839.

He was the eldest son of William Sleigh of Whitehall and Haworth House, County Middlesex, Counsellor-at-Law.

He was born 13th September 1781, and was rated as midshipman on board his uncle Sir James Wallace's ship. Later, he entered the army and saw service in Flanders 1795, being present at the actions in North Holland and the Helder, September and October 1799. He served in the Peninsular War and commanded the 11th Dragoons at the battle of Waterloo, towards the close of which the command of the 4th Brigade devolved upon him. He saw service in India and commanded the cavalry division at the siege of Bhurtpore 1825-26.

On the 24th August 1839 he was appointed to the Colonelcy of the regiment, which position he held till his death 5th February 1865.

He was promoted a Major-General 22nd July 1830, a Lieut.-General 23rd November 1841, and a General 26th June 1854. In 1856 he was appointed a K.C.B.

General Sir James Hope Grant, K.C.B.,

Appointed Colonel 6th February 1865.

Was the youngest son of Mr. Francis Grant of Kilgraston, Perthshire, and was born there 22nd July 1808.

He joined the regiment as Cornet in August 1826 at Glasgow, Lord Loughborough being his commanding officer, of whom he writes: "A better C.O. never lived, a straightforward, honest, fine fellow, with great judgment and talent."

Nearly forty years later (1865) Sir Hope was appointed to the Colonelcy of the regiment, and on writing to Lord Loughborough (then Lord Rosslyn) apologising for the "Cornet" having been preferred to the "Colonel," Lord Rossyln wrote: "Though I should have liked to have returned to the old Regiment, I felt that my claims could not stand in competition with yours, and that if the old Latin saying should be acted upon 'Detur digniori,' it could not have been given to any one but yourself, and I heartily congratulate you and the Regiment on the event."

In 1831, he was in the House of Commons with his brother-in-law Mr. James Lindsay, M.P., and by mistake took part in a division, and had his vote recorded. Luckily this was not found out, as the penalty was imprisonment and a fine of £500.

In 1837 he was employed as temporary A.D.C. to Lord Greenock, commanding the troops at Edinburgh.

In 1841 Lord Saltoun, who proceeded to China with a force to assist Sir Hugh Gough, appointed him his Brigade-Major.

Lord Saltoun was a great musician, so also was Sir Hope.

In October 1842 Lord Saltoun assumed command of all the troops, and he appointed Grant his Assistant-Adjutant-General. He was present at the assault and capture of Chin Kiang Foo, and at the landing before Nankin. Awarded a medal.

In 1842 the regiment was augmented, and Grant thus got his Majority without purchase.

In 1843 he was appointed a C.B.

In 1844 he joined the regiment at Cawnpore, and was present with it at the battle of Sabraon in 1846. He commanded the regiment at the battles of Chillianwallah and Goojerat.

In February 1847 he married Elizabeth Helen, daughter of Benjamin Tayler, a Sudder Judge.

In 1849 he was promoted a Brevet-Lieut.-Colonel for service during the Second Sikh War; and in 1850, owing to the sudden death of his next senior officer, Lieut.-Colonel Fullerton, he was appointed without purchase to the command of the regiment.

On the outbreak of the Mutiny he was appointed to command a cavalry brigade formed from the troops at Umballa and Meerut; he commanded the cavalry division at the siege and capture of Delhi, at which place on retreat of the rebels, he was in the first instance left in command; but later was appointed to replace Colonel Greathead in the command of a column marching to the relief of Lucknow.

He was appointed a Brigadier-General by Sir Colin Campbell, the Commander-in-Chief in India; and under him commanded the force which relieved Lucknow, and fought the battle of Cawnpore December 1857.

In 1858 he was appointed a Major-General, and a K.C.B. for his eminent services as General of the cavalry division at the siege and capture of Delhi, and the relief, siege, and capture of Lucknow.

During the latter operations he performed most excellent services and was mentioned many times in despatches. He was next placed in command of the force ordered to clear the province of Oude, and fought many engagements, concluding the campaign in June 1859.

Towards the end of the year he was promoted a Lieut.-General, and appointed to the command of a British force which was proceeding to China to co-operate with the French, which operations terminated in the capture of Pekin.

For this service Sir Hope received the thanks of Parliament, was appointed to the 2nd class of the Legion of Honour, and received two further clasps to his (1842) China medal.

In 1861 he was appointed a G.C.B. and given command of the army in Madras.

In July 1865 his services abroad ended, and he became Quartermaster-General to the army at the Horse Guards. He was this year appointed to the Colonelcy of the regiment.

In November 1870 he was selected to command the Aldershot division, and was instrumental in reviving the holding of autumn manœuvres. He was promoted a General in 1872, and died on the 7th March 1875, being buried at the Grange Cemetery, Edinburgh.

General Sir Archibald Little, G.C.B.

Appointed Colonel 8th March 1875.

General Little was the second son of Archibald Little, of Shabden Park, Surrey; which place he bought, after selling the old family property in Dumfriesshire. He was born in 1810, and was appointed to the regiment as Cornet by purchase on 4th October 1831. He was present with the regiment during the Sutlej campaign 1846 at the battle of Sabraon. Medal with clasp.

He obtained command of the regiment in March 1858. When the Mutiny broke out he was on leave in England, but on his return to India was at once appointed Brigadier in command of the cavalry at the "Relief of Lucknow," and commanded the Post at Dilkoosha from the 16th to 24th November 1857, for which he received the thanks of the Commander-in-Chief and the Governor-General in Council, and was appointed a C.B.

He commanded the cavalry and three troops Royal Horse Artillery, at the battle of Cawnpore and pursuit of the Gwalior rebels 6th December, including the actions of Serai Ghat and Khodargunge.

He commanded the 1st Cavalry Brigade at the siege of Lucknow until severely wounded through the left elbow.

Mentioned in despatches three times. Medal with two clasps.

He gave up the command of the regiment, in which he had served continuously, in May 1861. He was promoted a Major-General in March 1868, a Lieut.-General in July 1877, and a General in January 1880.

He was appointed Colonel of the regiment in March 1875, which appointment he held till his death in June 1891.

He was appointed a K.C.B. in June 1869, and a G.C.B. in 1890.

Lieut.-General Sir William Drysdale, K.C.B.

Appointed Colonel 11th June 1891.

He was born the 25th March 1819, and was appointed as Cornet, 4th Light Dragoons, 29th December 1835. He served with this regiment in the Afghan campaign 1839, under Lord Keane, being present at the siege and capture of Ghuznee, for which he was awarded a medal.

In April 1842 he was transferred to the 9th Lancers, and was present at the action of "Punniar," December 1843. He served in the Sutlej and Punjab campaigns, being present at the battle of Sabraon (medal); and taking part in the passage of the Chenab, in the action of Ramnuggar, and the battles of Goojerat and Chillianwallah. (Medal with two clasps.)

He served with the regiment throughout the Indian Mutiny, was present at Delhi during the siege operations, and commanded it from the 18th August to the fall of the city 20th September 1857, during which time he had his horse shot.

He commanded the regiment with Greathed's column at the action of Bolundshuhur, where his horse was shot and he wounded.

He was present at the actions of Allyghur, Agra, and the relief of Lucknow. For his services during this period he was mentioned in despatches, promoted Brevet-Lieut.-Colonel, awarded the C.B., and the medal with two clasps. These rewards however, only poorly record the excessive gallantry and soldier-like qualities of General Drysdale, who from all authentic accounts must have been the most efficient, reliable, gallant, and retiring officer; among many men of the same character at this period.

In May 1861, he succeeded to the command of the regiment vice Colonel Little, which position he held till May 1866.

He was promoted to the rank of Major-General January 1870, was appointed Colonel 18th Hussars in March 1885, and was transferred to the Colonelcy of the regiment in June 1891; on the death of his old brother officer, Sir Archibald Little.

He was promoted a Lieut.-General in July 1881, and was made a K.C.B. in 1894.

He died in August 1900, and was succeeded in the Colonelcy of the regiment by another old brother officer, Major-General H. A. Bushman, C.B.

He left to the commanding officer of the regiment, £2000 in trust, to distribute the income between "the best behaved non-commissioned officers and men; or partly between such officers and men, and partly in prizes for lance and sword exercises, to be competed for in such regiment."

MAJOR-GENERAL HENRY AUGUSTUS BUSHMAN, C.B.,

Appointed Colonel 8th August 1900.

Was born in June 1841, and was appointed to the regiment, 9th November 1858. On the 1st July 1859, he was transferred to the 7th Light Dragoons (7th Hussars), of which regiment he was Adjutant from October 1863 till March 1872, when he was reappointed to the 9th Lancers as Captain. He took part in the campaign on the North-West Frontier of India in 1863, in the Zulu campaign 1879, concluding operations and final operations against Sekukuni. Mentioned in despatches, medal with clasp, and brevet of Lieut.-Colonel.

In 1880 he succeeded to the command of the regiment vice Colonel Cleland, who had died of wounds. He was present with the regiment during the Afghan War 1879-80, taking part in the march from Kabul to Kandahar, and the battle of Kandahar 1st September 1880. Mentioned in despatches, medal with clasp and bronze star.

In May 1881 he was appointed a C.B., and was promoted to the rank of Major-General in 1895.

On the death of Lieut.-General Sir William Drysdale, K.C.B., in August 1900, he was appointed to the Colonelcy of the regiment.

ROLL OF THE COMMANDING OFFICERS OF THE REGIMENT.

1715-1904.

Name.		From	To
Col. and Lieut.-Col.	William Duncombe	24 July 1715	About Oct. 1729
Lieut.-Col.	Dallway	October 1729	March 1733
,,	Blakeney	March 1733	July 1737
,,	Francis Ligonier	8 July 1737	August 1739
,,	Daniel Degennes	August 1739	June 1753
,,	Francis Reynolds	June 1753	April 1756
,,	Owen Wynne	6 April 1756	16 June 1761
,,	John Wynne	17 June 1761	6 July 1766
,,	Arthur Preston	7 July 1766	24 June 1785
,,	Anthony Lefroy	25 June 1785	29 July 1791
,,	Joseph Westenra	30 July 1791	30 Mar. 1793
,,	The Hon. Brydges-Trecothic-Henniker	31 Mar. 1793	31 Dec. 1796
,,	The Hon. Thomas Mahon	1 Jan. 1797	23 Jan. 1811
,,	William Chabot	24 Jan. 1811	3 Mar. 1813
,,	Charles Morland	4 Mar. 1813	3 Mar. 1828
,,	Lord Loughborough	4 Mar. 1828	27 Jan. 1842
,,	C. Rowley	28 Jan. 1842	Retired at once
,,	Alexander Campbell, K.H., C.B.	28 Jan. 1842	23 Mar. 1850
,,	J. A. Fullerton, C.B.	24 Mar. 1850	28 April 1850
,,	J. Hope Grant, K.C.B.	29 April 1850	4 Mar. 1858
,,	Archibald Little, C.B.	5 Mar. 1858	23 May 1861
,,	William Drysdale, C.B.	24 May 1861	30 April 1866
,,	Charles P. Johnson	1 May 1866	18 May 1869
,,	The Hon. Ivo-de-Vesci W. Fiennes	19 May 1869	26 June 1874
,,	Henry Marshall	27 June 1874	26 June 1879
,,	Robert S. Cleland	27 June 1879	7 Aug. 1880
,,	Henry A. Bushman, C.B.	8 Aug. 1880	14 Dec. 1884
,,	Henry R. Abadie	15 Dec. 1884	1 July 1887
,,	The Hon. Heneage Legge	1 July 1887	15 Dec. 1890
,,	J. A. Stewart-Mackenzie	15 Dec. 1890	14 Dec. 1895
,,	Bloomfield Gough	15 Dec. 1895	14 Mar. 1900
,,	Malcolm O. Little	15 Mar. 1900	14 Mar. 1904
,,	The Hon. Claude Willoughby	15 Mar. 1904	

Colonel Charles Morland.

He was son of Thomas Morland of Court Lodge, Lamberhurst, Kent; and entered the army as a Cornet 12th Light Dragoons, 6th April 1796, rising to be Captain of a troop in the same regiment in November 1799.

He was promoted Major in the 17th Light Dragoons July 1809, and exchanged to the 9th Light Dragoons in February 1810.

He purchased his Lieut.-Colonelcy in the 50th Regiment, October 1812, and exchanged back into the regiment (with Lieut.-Colonel Chabot) as Lieut.-Colonel Commanding March 1813.

In January 1812 he married the Lady Caroline E. Courtenay, a daughter of the 9th Earl of Devon.

In May 1825 he was appointed an A.D.C. to his Majesty King George IV., with the rank of Colonel in the army.

On the 6th March 1828 owing to ill-health, he was obliged to relinquish the command of the regiment, which he had held for fourteen years.

His successors of all ranks in the "Ninth" have very much to thank him for.

It was he who established the excellent orders, system, and general good feeling among all ranks, which has worn, with all credit, materially unaltered to the present date. By his ability, zeal, and love of the regiment, he brought it to the high state of discipline and efficiency which, during his own period of command, was always most highly commented upon, and which, I think, it may be said, his successors as commanding officers, regimental officers, non-commissioned officers and men, have most fully, proudly, and devotedly upheld.

He died in Paris 14th June 1828.

LIEUT.-COLONEL JAMES ALEXANDER ST. CLAIR ERSKINE, Baron Loughborough of Loughborough, County Surrey (Earl of Rosslyn from 1837),

Was the eldest son of the full Colonel of the regiment, born February 1802, and appointed to the regiment as Cornet February 1819, with which he continued to serve in all ranks until he obtained the Lieut.-Colonelcy in March 1828. This position he held for fourteen years till January 1842, having during a period of peace carried on, perhaps further improved, the routine established by his predecessor.

He became a General in the army, was Colonel of the 7th Hussars from 1841 to 1846, was Master of the Buckhounds 1859, and Under Secretary of State for War to the Duke of Wellington.

ROLL OF THE ADJUTANTS.

Rank.	Name.	Remarks.	From	To
Lieut.	Edward Clayton	...	Oct. 1742	
,,	Thomas Crofton	...	Sept. 1746	Dec. 1749
,,	Henry Mann	...	12 Mar. 1754	26 Jan. 1768
,,	Thomas Henry Mann	...	26 Jan. 1768	12 July 1777
,,	Richard Johnston	...	12 July 1777	27 Oct. 1798
Cornet	Thomas Butterfield	...	27 Oct. 1798	April 1803
Lieut.	John Phillip Perry	Lieut. in the Army, 8 Nov. 1798	16 April 1803	26 Oct. 1804
Cornet	—— Hutchen	...	26 Oct. 1804	1806
,,	John Knott	...	30 Oct. 1806	4 July 1811
,,	John Wright	...	4 July 1811	27 April 1827
,,	Robert Cooke	Private 9th L. 1804. Commission 27 April 1827	28 April 1827	22 April 1842
,,	William Hamilton	Private 9th L. Feb. 1833. Commission 29 April 1842	29 April 1842	19 Nov. 1856
Lieut.	F. R. C. Grant	...	20 Nov. 1856	16 Mar. 1858
Cornet	R. Freer Thonger	Private 9th L. Feb. 1841. Commission 27 June 1857	16 Mar. 1858	29 June 1865
Lieut.	B. Cave Browne-Cave	...	30 June 1865	30 April 1866
,,	William Savile	...	1 May 1866	19 April 1867
,,	F. Grissell	...	20 April 1867	19 May 1869
Cornet	William Percy	Private 4th L. Dragoons, 12 April 1842. Commission 12th L. 14 Aug. 1867. Transferred to 9th L. same date	19 May 1869	24 July 1875
Lieut.	The Hon. Eustace Vesey	...	24 July 1875	22 June 1877
,,	E. B. McInnis	Private 16th L. 16 Mar. 1865. Commission 21st Hussars 29 Apr. 1874, 9th L. 23 Oct. 1875	22 June 1877	17 Nov. 1882
Major	J. J. Scott Chisholme	...	18 Nov. 1882	14 Dec. 1884
Lieut.	F. H. Reynard	...	20 May 1885	6 Aug. 1886
Captain	F. H. Reynard	Re-appointed on exchange from 7th Hussars	20 Oct. 1886	29 July 1887
Lieut.	W. K. Jenner	...	30 July 1887	29 July 1891
,,	The Hon. Victor Pelham	...	30 July 1891	30 Aug. 1892
,,	The Hon. L. King-Noel	...	31 Aug. 1892	12 Mar. 1895
,,	D. J. Beale-Browne	...	13 Mar. 1895	19 April 1898
,,	E. F. Bell	...	20 April 1898	19 April 1902
,,	J. Gordon-Stirling	...	20 April 1902	22 May 1902
Bt.-Maj.	D. G. Campbell	...	6 Sept. 1902	27 April 1904
Captain	H. M. Durand	...	28 April 1904	

ROLL OF THE QUARTERMASTERS.

Name.	Remarks.	From	To
Richard Sidley . .	...	June 1800	1809
Thomas Heley . .	...	2 Nov. 1809	11 Aug. 1825
Thomas Wright . .	Private 9th L. 1805	11 Aug. 1825	17 Aug. 1838
J. C. Williamson . .	Private 9th L. 1810	17 Aug. 1838	17 April 1841
John Heydon . . .	Private 9th L. 1807	30 April 1841	13 May 1842
Andrew Allan . .	Private Royal Horse Guards 1824	13 May 1842	12 Nov. 1852
Peter House . . .	Private 9th L. 1840	12 Nov. 1852	24 May 1861
Samuel Seggie . .	...	24 May 1861	29 Oct. 1869
George Fuller. . .	Private 9th L. 1843	30 Oct. 1869	26 Sept. 1879
John Forsythe . .	Private 9th L. 1858	27 Sept. 1879	22 Dec. 1893
David Laing . . .	To half-pay—ill-health	23 Dec. 1897	21 Dec. 1903
W Grant . . .	Orderly - room Sergeant and Regt.-Serg.-Major of the Regt. previously.	22 Dec. 1903	

APPENDIX VII.

ANNUAL ARMY LISTS.

MAJOR-GENERAL WYNNE'S DRAGOONS

Rank.	1715	Date of Commission.	1717	Date of Commission.
Colonel	Owen Wynne		Owen Wynne (Maj.-General)	22 July 1715
Lieut.-Col.	H. Pearson		William Duncombe (Colonel)	24 July 1715
Major	John Dunbar		John Dunbar	22 July 1715
Captain	John, Lord Leslie		Andrew Knox	,,
,,	Andrew Knox		Henry Smith	,,
,,	Henry Smith		John, Lord Leslie (later the Earl of Rothes)	,,
Capt. and Lieut.	H. Crawford		Henry Crawford	,,
Lieutenant	Wm. Witherington		John Warnes	,,
,,	Edward Whitney		William Humphreys	,,
,,	William Humphreys		Edward Whitney	,,
,,	Jacob Warnes		Owen Wynne	20 Jan. 1716
,,	Gustavus Hamilton		Wm. Witherington	22 July 1715
Cornet	Owen Wynne		William Carleton	,,
,,	William Carleton		Christopher Adams	,,
,,	Lewis Folliott		Roger Pemberton	,,
,,	James Hill		Lewis Folliott	,,
,,	Roger Pemberton		John Folliott	20 Jan. 1716
,,	Christopher Adams		John Hill	22 July 1715
Chaplain	..		Nicholas Milly	,,
Surgeon	..		Theodore Amyand	,,

AGENT—Francis Cocksedge.

NINTH REGIMENT OF DRAGOONS

Rank.	1727	Date of Commission.	1734	Date of Commission.	1737	Date of Commission.
Colonel	James Crofts		The Viscount Molesworth		Brig.-Gen. Sir John Cope	
Lieut.-Col.	William Duncombe		—— Blakeney		Francis Ligonier	
Major	Henry Goddard		Francis Ligonier		P. Malide	
Captain	Francis Ligonier		Paul Malide		Lord Charles Hay	
,,	Paul Malide		Lord Charles Hay		H. Montgomery	
,,	Henry Crawford		Hugh Montgomery		W. Humphreys	
Capt. and Lieut.	William Humphreys		William Humphreys		Owen Wynne	
Lieutenant	Francis Reynolds		Owen Wynne		William Carleton	
,,	Owen Wynne		William Carleton		James Hill	
,,	William Carleton		James Hill		Francis Reynolds	
,,	James Hill		William Green		Edward Webster	
,,	William Green		Francis Reynolds		Sir W. Costwick, Bt.	
Cornet	Edward Webster		Edward Webster		Thomas Aston	
,,	Sir W. Costwick, Bt.		Sir W. Costwick, Bt.		James Taylor	
,,	Thomas Aston		Thomas Aston		Edward Clayton	
,,	James Taylor		James Taylor		Edward Lee (Leigh)?	
,,	Edward Clayton		Edward Clayton		Demetrius James	
,,	Nathaniel Preston		—— Amherst		Thomas Darby	
Chaplain	Nicholas Milly		Nicholas Milly		Nicholas Milly	
Surgeon	Robert Archer		William Prince		William Prince	
Quarter-masters	..		Hodgson, Moore, Johnson, Crofton, Bermingham, Mercier		Cusack, Clark, Crofton, Jackson, Donellan, Howard	

Rank.	1739	Date of Commission.	1742	Date of Commission.	1746	Date of Commission.
Colonel	Brig.-Gen. Sir John Cope		Brig.-Gen. Sir John Brown		Henry de Grangue	
Lieut.-Col.	J. D. Degennes		J. D. Degennes		J. D. Degennes	
Major	P. Malide		P. Malide		P. Malide	
Captain	Lord Charles Hay		Lord C. Hay		W. Humphreys	
,,	H. Montgomery		W. Humphreys		O. Wynne	
,,	W. Humphreys		O. Wynne		F. Reynolds	
Capt. and Lieut.	Owen Wynne		F. Reynolds		T. Babington	
Lieutenant	W. Carleton		E. Webster		E. Clayton	
,,	F. Reynolds		Sir W. Costwick, Bt.		H. Westenra	
,,	E. Webster		E. Clayton		T. Darby	
,,	Sir W. Costwick, Bt.		H. Westenra		H. Clarke	
,,	R. Fitzgerald		Arthur Blennerhasset		T. Crofton	
Cornet	J. Taylor		H. Clarke		J. Cullen	
,,	E. Clayton		R. Pringle		B. Barnes	
,,	D. James		A. Rollo		H. Moore	
,,	Thomas Darby		E. Ellis		J. Moutray	
,,	Henry Clarke		J. Cullen		St. Leger Aldworth	
,,	R. Pringle		St. Leger Aldworth		R. French	
Chaplain	Nicholas Milly		Nicholas Milly		N. Milly	
Surgeon	William Prince		W. Prince		W. Prince	
Adjutant	..		Edward Clayton		T. Crofton	

NINTH REGIMENT OF DRAGOONS—*continued*

Rank.	1749[1]	Date of Commission.	1751	Date of Commission.	1754	Date of Commission.
Colonel	Henry de Grangue		George Reade		George Reade	Nov. 1747
Lieut.-Col.	J. D. Degennes		J. D. Degennes		Francis Reynolds	June 1753
Major	..		Francis Reynolds		Owen Wynne	,,
Captain	Francis Reynolds		Owen Wynne		Edward Clayton	Aug. 1747
,,	Owen Wynne		E. Clayton		W. Lushington	Apr. 1749
,,	Edward Clayton		W. Lushington		H. Westenra	June 1753
,,	William Lushington		..		..	
Capt. and Lieut.	..		H. Westenra		H. Clarke	,,
Lieutenant	H. Clarke		H. Clarke		J. Cullen	Aug. 1747
,,	J. Cullen		R. French		Hugh Moore	Nov. 1747
,,	H. Westenra		J. Cullen		Robert French	Apr. 1750
,,	H. Moore		Hugh Moore		Bartholomew Barnes	June 1753
,,	St. Leger Aldworth		St. L. Aldworth		Paul Bush	,,
Cornet	B. Barnes		B. Barnes		W. Latham	Jan. 1750
,,	Thomas Desbrisay		Sir C. Bingham, Bt.		Sir C. Bingham, Bt.	Apr. 1750
,,	J. Moutray		W. Latham		Thomas Hay	June 1751
,,	Christopher Daulhat		Paul Bush		James Brown	Jan. 1752
,,	R. French		Thomas Hay		Dixie Coddington	June 1753
,,	..		E. Loftus		H. Hughes	,,
Chaplain	George Preston		Thomas Maddox		Thomas Madden	Aug. 1751
Surgeon	William Prince		William Prince		T. Tibboe	Mar. 1754
Adjutant	..		..		H. Mann	,,

[1] In this year no Muster Roll for the major's troop is to be found.

Rank.	1756	Date of Commission.	1758	Date of Commission.	1759 and 1760	Date of Commission.
Colonel	Philip Honywood	May 1756	P. Honywood		Henry Whitley	6 April 1759
Lieut.-Col.	Owen Wynne	April 1756	Owen Wynne		Owen Wynne	
Major	Edward Clayton	,,	E. Clayton		Edward Clayton	
Captain	H. Westenra		J. Cullen		John Cullen	
,,	J. Cullen	April 1756	H. Clarke		Henry Clarke	
,,	Henry Clarke	Nov. 1756	R. French	Nov. 1757	Robert French	
Capt. and Lieut.	Hugh Moore	,,	H. Moore		Hugh Moore	
Lieutenant	Robert French		P. Bush		Paul Bush	
,,	B. Barnes		W. Latham		D. Coddington	
,,	P. Bush		J. Brown		Henry Hughes	
,,	W. Latham	April 1756	D. Coddington	July 1757	Richard Saunders	7 Feb. 1759
,,	James Brown	Nov. 1756	H. Hughes	Nov. 1757	Robert Waller	3 April 1759
Cornet	D. Coddington		R. Saunders		John Francis Erskine	
,,	Henry Hughes		E. Sterne		Richard Vandeleur	
,,	Richard Saunders	April 1756	R. Waller		John Adlercron	
,,	Richard J. Powis	,,	John F. Erskine	July 1757	Henry Mann	7 Feb. 1759
,,	Enoch Sterne	,,	Richard Vandeleur	,,	Arthur Waller	,,
,,	Robert Waller	Nov. 1756	John Adlercron	Nov. 1757	Mathew Lyster	3 April 1759
Chaplain	T. Madden		Maurice Gough	April 1758	Maurice Gough	
Adjutant	H. Mann		H. Mann		H. Mann	
Surgeon	T. Tibboe		T. Tibboe		T. Tibboe	

AGENT—Mr. Chaigneau, Dublin Castle.

NINTH REGIMENT OF DRAGOONS—*continued*

Rank.	1761	Date of Commission.	1762	Date of Commission.	1763	Date of Commission.
Colonel	Henry Whitley		Henry Whitley		Henry Whitley	
Lieut.-Col.	Owen Wynne		John Wynne (Army, 7 Dec. 1759)	17 June 1761	John Wynne	
Major	Henry Clarke	29 Aug. 1760	Henry Clarke		Henry Clarke	
Captain	John Cullen		Robert French		Robert French	
,,	Robert French		Robert Waller		Robert Waller	
,,	Robert Waller	29 Aug. 1760	William Hull (Army, 18 Oct. 1758)	17 Mar. 1761	Richard Vandeleur	12 Feb. 1762
Capt. and Lieut.	D. Coddington	22 Mar. 1760	D. Coddington		D. Coddington	
Lieutenant	Paul Bush		Paul Bush		Paul Bush	
,,	Henry Hughes		Henry Hughes		Richard Saunders	
,,	Richard Saunders		Richard Saunders		Henry Mann	17 Sept. 1761
,,	Richard Vandeleur	22 Mar. 1760	Richard Vandeleur		Mathew Lyster	1 Apr. 1762
,,	John Adlercron	29 Aug. 1760	John Adlercron			
Cornet	Henry Mann		Henry Mann		Anthony Scott	
,,	Arthur Waller		Arthur Waller		Thomas Ormsby	
,,	Mathew Lyster		Mathew Lyster		Samuel Lawson	
,,	Anthony Scott	17 Jan. 1760	Anthony Scott		Francis Lodge	17 Sept. 1761
,,	Thomas Ormsby	22 Mar. 1760	Thomas Ormsby		William Adlercron	1 April 1762
,,	Samuel Lawson	29 Aug. 1760	Samuel Lawson		Barth. Mahon	30 July 1762
Chaplain	Maurice Gough		Maurice Gough		Maurice Gough	
Adjutant	Henry Mann		Henry Mann		Henry Mann	
Surgeon	Thomas Tibboe		Thomas Tibboe		Thomas Tibboe	

Rank.	1764 and 1765	Date of Commission.	1766	Date of Commission.	1767	Date of Commission.
Colonel	Henry Whitley		H. Whitley (Maj.-Gen., 13 Aug. 1761)		Henry Whitley	
Lieut.-Col.	John Wynne		John Wynne		Arthur Preston	7 July 1766
Major	Robert French	11 Oct. 1762	Robert French		Robert Waller	1 Oct. 1766
Captain	Robert Waller		Robert Waller		Richard Saunders	
,,	Richard Vandeleur		Richard Saunders	7 Dec. 1764	Thomas Ormsby	
,,	John Staples	11 Oct. 1762	Thomas Ormsby	14 Aug. 1765	Rich. Earl of Barrymore	1 Oct. 1766
Capt. and Lieut.	D. Coddington		D. Coddington		Paul Bush	18 June 1766
Lieutenant	Paul Bush		Paul Bush		Henry Mann	
,,	Richard Saunders		Henry Mann		Mathew Lyster	
,,	Henry Mann		Mathew Lyster		Samuel Lawson	
,,	Mathew Lyster		Samuel Lawson	7 Dec. 1764	Francis Lodge	
,,	Thomas Ormsby	1 April 1762	Francis Lodge	14 Aug. 1765	William Adlercron	18 June 1766
Cornet	Anthony Scott		Anthony Scott		George Marley	
,,	Samuel Lawson		William Adlercron		Barth. Mahon	
,,	Francis Lodge		George Marley		James Garden	
,,	William Adlercron		Barth. Mahon		Miles Strickland	18 April 1766
,,	George Marley	1 April 1762	James Garden	7 Dec. 1764	Thomas Brown	16 May 1766
,,	Barth. Mahon		Marcus Browne	14 Aug. 1765	William Ormsby	18 June 1766
Chaplain	Maurice Gough		Maurice Gough		Maurice Gough	
Adjutant	Henry Mann		Henry Mann		Henry Mann	
Surgeon	Thomas Tibboe		Andrew M'Dowall	13 Feb. 1765	Andrew M'Dowall	

AGENTS—William and John Chaigneau, Dublin Castle.

NINTH REGIMENT OF DRAGOONS—*continued*

Rank.	1768	Date of Commission.	1769	Date of Commission.	1770	Date of Commission.
Colonel	Henry Whitley		Henry Whitley		Henry Whitley	
Lieut.-Col.	Arthur Preston		Arthur Preston		Arthur Preston	
Major	Robert Waller		Robert Waller		Robert Waller	
Captain	Richard Saunders		Richard Saunders		Richard Saunders	
,,	Thomas Ormsby		Thomas Ormsby		Thomas Ormsby	
,,	Rich. Earl of Barrymore		Rich. E. of Barrymore		Mathew Lyster	1 Jan. 1769
Capt. and Lieut.	Mathew Lyster	27 May 1767	Mathew Lyster		Samuel Lawson	1 Jan. 1769
Lieutenant	Henry Mann		Henry Mann		Henry Mann	
,,	Samuel Lawson		Samuel Lawson		Francis Lodge	
,,	Francis Lodge		Francis Lodge		William Adlercron	
,,	William Adlercron		William Adlercron		George Marley	
,,	George Marley	27 May 1767	George Marley		James Garden	1 Jan. 1769
Cornet	Barth. Mahon		James Garden		Miles Strickland	
,,	James Garden		Miles Strickland		William Ormsby	
,,	Miles Strickland		Thomas Brown		John Blake	
,,	Thomas Brown		William Ormsby		Barton Lodge	
,,	William Ormsby		John Blake		Nathaniel Preston	1 Jan. 1769
,,	John Blake	27 May 1767	Barton Lodge	2 Mar. 1768	Thomas Curtis	11 Oct. 1769
Chaplain	Maurice Gough		Maurice Gough		Maurice Gough	
Adjutant	Thomas Henry Mann	26 Jany. 1768	Thomas Henry Mann		Thomas Henry Mann	
Surgeon	Andrew M'Dowall		Andrew M'Dowall		Andrew M'Dowall	

Rank.	1771	Date of Commission.	1772	Date of Commission.	1773	Date of Commission.
Colonel	Henry Whitley (Lt.-Gen., 30 Apr. 1770)		Jas. Johnston (Maj.-Gen., 30 April 1770)	17 Jan. 1771	James Johnston	
Lieut.-Col.	Arthur Preston		Arthur Preston		Arthur Preston	
Major	Robert Waller		Robert Waller		Robert Waller	
Captain	Richard Saunders		Richard Saunders		Richard Saunders	
,,	Thomas Ormsby		William Adlercron	30 April 1771	William Adlercron	
,,	Mathew Lyster		Marcus Anto. Tuite	30 April 1771	Marcus Anto. Tuite	
Capt. and Lieut.	Henry Mann	2 June 1770	Henry Mann		Henry Mann (Capt., 25 May 1772)	
Lieutenant	Francis Lodge		Francis Lodge		Francis Lodge	
,,	William Adlercron		James Garden		James Garden	
,,	James Garden		William Ormsby		William Ormsby	
,,	William Ormsby	2 June 1770	John Blake	30 April 1771	John Blake	
,,	W. J. Skeffington	16 Aug. 1770	Toby Cauldfield	5 June 1771	Barton Lodge	14 Mar. 1772
Cornet	John Blake		Barton Lodge		Nathaniel Preston	
,,	Barton Lodge		Nathaniel Preston		Thomas Curtis	
,,	Nathaniel Preston		Thomas Curtis		Tennison Smith	
,,	Thomas Curtis		The Hon. Henry Hewitt		Wm. Stone Montgomery	
,,	The Hon. Henry Hewitt	2 June 1770	Tennison Smith	30 April 1771	John Beevor	14 Mar. 1772
,,	G. Lord Ferrers	29 Sept. 1770	Wm. Stone Montgomery	16 Dec. 1771	The Hon. Arthur Cole (Lieut., 23 June 1769)	26 Sept. 1772
Chaplain	Maurice Gough		Maurice Gough		Maurice Gough	
Adjutant	Thomas Henry Mann		Thomas Henry Mann		Thomas Henry Mann	
Surgeon	Andrew M'Dowall		Andrew M'Dowall		Andrew M'Dowall	

AGENT—Wm. Montgomery, Dublin.

NINTH REGIMENT OF DRAGOONS—*continued*

Rank.	1774	Date of Commission.	1775	Date of Commission.	1776	Date of Commission.
Colonel	James Johnston		James Johnston		Flower Mocher (Col. 25 May 1772	27 Apr. 1775
Lieut.-Col.	Arthur Preston (Army, 5 Jan. 1762)		Arthur Preston		Arthur Preston	
Major	Robert Waller		Richard Saunders	12 Mar. 1774	Richard Saunders	
Captain	Richard Saunders		William Adlercron		William Adlercron	
,,	William Adlercron		Marcus Anto. Tuite		Marcus Anto. Tuite	
,,	Marcus Anto. Tuite		Nathaniel Preston	12 Mar. 1774	Nathaniel Preston	
Capt. and Lieut.	Henry Mann		Henry Mann		Arthur Pomeroy	24 Feb. 1775
Lieutenant	Francis Lodge		Francis Lodge		James Garden	
,,	James Garden		James Garden		William Ormsby	
,,	William Ormsby		William Ormsby		Barton Lodge	
,,	Barton Lodge		Barton Lodge		John Beevor	
,,	Nathaniel Preston	6 Nov. 1772	John Beevor	12 Mar. 1774	Anderson Saunders	1 May 1775
Cornet	Thomas Curtis		Wm. Stone Montgomery		Hon. Arthur Cole	
,,	Wm. Stone Montgomery		Hon. Arthur Cole		John Langley	
,,	John Beevor		John Langley		John Craven Carden	
,,	The Hon. Arthur Cole		Anderson Saunders		Richard Donaldson	
,,	John Langley	6 Nov. 1772	John Craven Carden	12 Mar. 1774	Robert Kindillon	20 Mar. 1775
,,	Anderson Saunders	6 Nov. 1772	Richard Donaldson	16 July 1774	Thomas Henry Mann	1 May 1775
Chaplain	Anthony Armstrong	20 Feb. 1773	Anthony Armstrong		Anthony Armstrong	
Adjutant	Thomas Henry Mann		Thomas Henry Mann		Thomas Henry Mann	
Surgeon	Andrew M'Dowall		Andrew M'Dowall		Andrew M'Dowall	

AGENT—Sir Wm. Montgomery, Bt., Dublin.

Rank.	1777	Date of Commission.	1778	Date of Commission.	1779	Date of Commission.
Colonel	Flower Mocher		F. Mocher (Maj.-Gen., 29 Aug. 1777)		Flower Mocher	
Lieut.-Col.	Arthur Preston		Arthur Preston (Col., 29 Aug. 1777)		Arthur Preston	
Major	Richard Saunders		Richard Saunders		Arthur Pomeroy	14 Feb. 1779
Captain	William Adlercron		William Adlercron		William Adlercron	
,,	Marcus Anto. Tuite		Marcus Anto. Tuite		Nathaniel Preston	
,,	Nathaniel Preston		Nathaniel Preston		George Warde	20 May 1778
Capt. and Lieut.	Arthur Pomeroy		Tho. Howe Paliser	5 Oct. 1776	Tho. Howe Paliser	
Lieutenant	James Garden		James Garden		James Garden	
,,	William Ormsby		Barton Lodge		Barton Lodge	
,,	Barton Lodge		John Beevor		John Beevor	
,,	John Beevor		Anderson Saunders		Anderson Saunders	
,,	Anderson Saunders		Arthur Carter	14 Jan. 1777	Richard Donaldson	5 Feb. 1778
Cornet	Hon Arthur Cole		Richard Donaldson		Robert Kindillon	
,,	John Langley		Robert Kindillon		James Colthurst	
,,	Richard Donaldson		Robert Riddell		Richard Gabbett	
,,	Robert Kindillon		James Colthurst	12 May 1777	Arthur Innes	
,,	Thomas Henry Mann		Richard Gabbett	12 July 1777		
,,	Robert Riddell	16 Dec. 1775	Anthony Innes	22 July 1777		
Chaplain	Anthony Armstrong		Anthony Armstrong		Anthony Armstrong	
Adjutant	Thomas Henry Mann		Richard Johnston	12 July 1777	Richard Johnston	
Surgeon	Andrew M'Dowall		Andrew M'Dowall		Andrew M'Dowall	

NINTH REGIMENT OF DRAGOONS—*continued*

Rank.	1780	Date of Commission.	1781	Date of Commission.	1782	Date of Commission.
Colonel	Flower Mocher		Flower Mocher		Flower Mocher	
Lieut.-Col.	Arthur Preston		Arthur Preston		Arthur Preston (Maj.-Gen., 19 Oct. 1781)	
Major	Arthur Pomeroy		Arthur Pomeroy		Arthur Pomeroy	
Captain	Nathaniel Preston		Nathaniel Preston		Arthur Ormsby	
,,	George Warde		Arthur Ormsby		Anderson Saunders	10 Dec. 1781
,,	Arthur Ormsby (Army, 20 Oct. 1774)	7 Apr. 1780	Andrew Corbet (Army, 12 Dec. 1777)	10 Nov. 1780		
Capt. and Lieut.	Thos. Howe Paliser		Thos. Howe Paliser		Thos. Howe Paliser	
Lieutenant	Barton Lodge		Barton Lodge		Barton Lodge	
,,	John Beevor		John Beevor		John Beevor	
,,	Anderson Saunders		Anderson Saunders		Richard Donaldson	
,,	Richard Donaldson		Richard Donaldson		Arthur Innes	10 Dec. 1781
,,	The Hon. George Walpole	17 Apr. 1780	The Hon. George Walpole			
Cornet	Robert Kindillon		Robert Kindillon		Robert Kindillon	
,,	James Colthurst		James Colthurst		James Colthurst	
,,	Arthur Innes		Arthur Innes		—— Coen	
,,	—— Coen	12 Aug. 1779	—— Coen		Richard Johnston	
,,	Mitchelburne Knox (Army, 3 June 1778)	24 Dec. 1779	Richard Johnston	9 Dec. 1780	Philip Ormsby	
,,			Philip Ormsby	26 May 1781	James Russell	10 Dec. 1781
Chaplain	Anthony Armstrong		Anthony Armstrong		Anthony Armstrong	
Adjutant	Richard Johnston		Richard Johnston		Richard Johnston	
Surgeon	Andrew M'Dowall		Andrew M'Dowall		Andrew M'Dowall	

NINTH LIGHT DRAGOONS

Rank.	1783	Date of Commission.	1784	Date of Commission.	1785	Date of Commission.
Colonel	Flower Mocher (Lt.-Gen., 20 Nov. 1782)		Flower Mocher		Flower Mocher	
Lieut.-Col.	Arthur Preston		Arthur Preston		Arthur Preston	
Major	Arthur Pomeroy		Arthur Pomeroy		Arthur Ormsby (Army, 19 Mar. 1783)	
Captain	Arthur Ormsby		Arthur Ormsby		Anderson Saunders	
,,	Anderson Saunders		Anderson Saunders		Mark Kerr	1 Oct. 1784
,,	Richard R. Wilford (Army, 26 March 1778)	20 Aug. 1782	William Hunt	24 May 1783	Humphrey Denis	28 Dec. 1784
Capt. and Lieut.	Thos. Howe Paliser		Thos. Howe Paliser		Thos. Howe Paliser	
Lieutenant	John Beevor		John Beevor		John Beevor	
,,	Richard Donaldson		Richard Donaldson		Richard Donaldson	
,,	Arthur Innes		Robert Kindillon		Robert Kindillon	
,,	Robert Kindillon	10 Dec. 1781	Arthur Innes		Arthur Innes	
,,	Richard Johnston	29 June 1782	Richard Johnston		Richard Johnston	
Cornet	James Colthurst		James Colthurst		James Colthurst	
,,	Philip Ormsby		Philip Ormsby		Philip Ormsby	
,,	James Russell		James Russell		John Browne	
,,	John Browne	13 April 1782	John Browne		James Hall	
,,	James Hall	29 June 1782	James Hall		Michael Fenton	29 July 1784
,,					Josiah Caldwell	28 Dec. 1784
Chaplain	Anthony Armstrong		Anthony Armstrong		Anthony Armstrong	
Adjutant	Richard Johnston		Richard Johnston		Richard Johnston	
Surgeon	Andrew M'Dowall		Andrew M'Dowall		Andrew M'Dowall	

NINTH LIGHT DRAGOONS—*continued*

Rank.	1786	Date of Commission.	1787	Date of Commission.	1788	Date of Commission.
Colonel	Flower Mocher		Flower Mocher		Flower Mocher	
Lieut.-Col.	Anthony Lefroy	25 June 1785	Anthony Lefroy		Anthony Lefroy	
Major	Arthur Ormsby		Arthur Ormsby		Arthur Ormsby	
Captain	Anderson Saunders		Anderson Saunders		Anderson Saunders	
,,	Mark Kerr		Mark Kerr		Mark Kerr	
,,	Humphrey Denis		Humphrey Denis		Humphrey Denis	
Capt. and Lieut.	Rich. Donaldson	19 Mar. 1785	Rich. Donaldson		Rich. Donaldson	
Lieutenant	John Beevor		John Beevor		John Beevor	
,,	Robert Kindillion		Robert Kindillon		Robert Kindillon	
,,	Arthur Innes		Arthur Innes		Arthur Innes	
,,	Richard Johnston		Richard Johnston		Richard Johnston	
,,	James Colthurst	19 Mar. 1785	James Colthurst		Philip Ormsby	30 June 1787
Cornet	Philip Ormsby		Philip Ormsby		J. Browne	
,,	John Browne		J. Browne		James Hall	
,,	James Hall		James Hall		Hyacinth Daly	
,,	Michael Fenton		Josiah Caldwell		Alfred F. Armstrong	31 Mar. 1787
,,	Josiah Caldwell		John Caldwell		Gustavus Hippersley	20 Dec. 1787
,,	John Caldwell	3 Dec. 1785	Hyacinth Daly (Army, 13 Dec. 1783)	16 Sept. 1786		
Chaplain	John Fortescue	25 June 1785	John Fortescue		John Fortescue	
Adjutant	Richard Johnston		Richard Johnston		Richard Johnston	
Surgeon	Andrew M'Dowall		Andrew M'Dowall		Andrew M'Dowall	

Rank.	1789	Date of Commission.	1790	Date of Commission.	1791	Date of Commission.
Colonel	Flower Mocher		Flower Mocher		Flower Mocher	
Lieut.-Col.	Anthony Lefroy		Anthony Lefroy		Anthony Lefroy	
Major	Joseph Westenra	31 Mar. 1788	Joseph Westenra		Joseph Westenra	
Captain	Anderson Saunders		Anderson Saunders		Anderson Saunders	
,,	Mark Kerr		Mark Kerr		Mark Kerr	
,,	Humphrey Denis		Humphrey Denis		Humphrey Denis	
Capt. and Lieut.	Rich. Donaldson		Rich. Donaldson		Rich. Donaldson	
Lieutenant	John Beevor		John Beevor		John Beevor	
,,	Robert Kindillon		Robert Kindillon		Arthur Innes	
,,	Arthur Innes		Arthur Innes		Richard Johnston	
,,	Richard Johnston		Richard Johnston		James Hall	31 May 1790
,,	Philip Ormsby		Philip Ormsby		William Gale	31 Dec. 1790
Cornet	John Browne		John Browne		Alfred F. Armstrong	
,,	James Hall		James Hall		John West Wilkins (Army, 5 Aug. 1783)	31 Aug. 1790
,,	Hyacinth Daly		Hyacinth Daly		Henry Loftus	30 Sept. 1790
,,	Alfred F. Armstrong		Alfred F. Armstrong		Wm. Jevers (Army, 25 Jan. 1788)	30 Nov. 1790
,,	Gustavus Hippersley		Gustavus Hippersley		Rich. Uniacke (Army, 28 Feb. 1788)	30 Nov. 1790
,,	Henry Cole (2 Lieut., 25 Sept. 1787)	31 Jan. 1788	John Wright	14 Mar. 1789	Robert John Love	31 Dec. 1790
Chaplain	Vere Essex Quaile	28 Feb. 1788	James Saurin	31 Aug. 1788	James Saurin	
Adjutant	Richard Johnston		Richard Johnston		Richard Johnston	
Surgeon	Andrew M'Dowall		Andrew M'Dowall		Andrew M'Dowall	

AGENTS—Messrs. Fraser and Reed, Dublin

NINTH LIGHT DRAGOONS—*continued.*

Rank.	1792	Date of Commission.	1793	Date of Commission.	1794	Date of Commission.
Colonel	Flower Mocher		Flower Mocher		Flower Mocher	
Lieut.-Col.	Joseph Westenra	30 July 1791	Joseph Westenra		B. Trecothick Henniker	31 Mar. 1793
Major	Anderson Saunders	,,	Anderson Saunders		Anderson Saunders	
Captain	Humphrey Denis		Humphrey Denis		Humphrey Denis	
,,	Ed. Witherington	,,	Ed. Witherington		Ed. Witherington	
,,	Andrew Philip Skene (Army, 19 Mar. 1783)	31 Aug. 1791	James Hall	31 July 1792	James Hall	
Capt. and Lieut.	Rich. Donaldson		Rich. Donaldson		Rich. Donaldson	
Lieutenant	John Beevor		John Beevor		John Beevor	
,,	Arthur Innes		Arthur Innes		Arthur Innes	
,,	Richard Johnston		Richard Johnston		Richard Johnston	
,,	James Hall		William Gale		William Gale	
,,	William Gale		Alfred F. Armstrong	,,	Alfred F. Armstrong	
Cornet	Alfred F. Armstrong		John West Wilkins		John West Wilkins	
,,	John West Wilkins		Henry Loftus		Henry Loftus	
,,	Henry Loftus		William Jevers		Robert John Love	
,,	William Jevers		Richard Uniacke		Lancelot Gubbings	31 Dec. 1792
,,	Richard Uniacke		Robert John Love		Francis Talbot	30 April 1793
,,	Robert John Love					
Chaplain	James Saurin		Marcus M'Causland	30 Nov. 1792	Marcus M'Causland	
Adjutant	Richard Johnston		Richard Johnston		Richard Johnston	
Surgeon	William Wallace	11 May 1791	William Wallace		William Wallace	

Rank.	1795	Date of Commission.	1796	Date of Commission.	1797	Date of Commission.
Colonel	Flower Mocher		Flower Mocher		Flower Mocher (Gen., 3 May 1796)	
Lieut.-Col.	B. Trecothick Henniker		B. Trecothick Henniker		B. Trecothick Henniker (Col., 3 May 1796)	
Major	Anderson Saunders		Anderson Saunders (Lieut.-Col., 13 Apr. 1795)		Anderson Saunders	
Captain	Humphrey Denis		Humphrey Denis (Maj., 6 May 1795)		Humphrey Denis	
,,	Ed. Witherington		Richard Donaldson (Maj., 6 May 1795)	19 Mar. 1785	Richard Donaldson	
,,	John Beevor	19 Dec. 1793	Ed. Witherington		Ed. Witherington	
,,	Arthur Innes	,,	John Beevor		John Beevor	
,,	..		Arthur Innes		Arthur Innes	
,,	..		The Hon. William J. Erskine (Army, 8 Nov. 1793)	30 Oct. 1794	The Hon. William J. Erskine	
Capt. and Lieut.	Richard Donaldson		Richard Johnston	28 Feb. 1795	Richard Johnston	
Lieutenant	Richard Johnston		Alfred F. Armstrong		Alfred F. Armstrong	
,,	William Gale		John West Wilkins		John West Wilkins	
,,	Alfred F. Armstrong		Henry Loftus		Henry Loftus	
,,	James Kearnay	9 May 1794	Robert John Love		Robert John Love	
,,	C. C. Patrickson (Army, 25 Mar. 1794)	14 June 1794	C. C. Patrickson		C. C. Patrickson	
,,	John West Wilkins		J. Lewis Higgins	6 Mar. 1795	J. Lewis Higgins	
,,	Henry Loftus		William Martin	3 July 1795	William Martin	
,,	Robert John Love		John Mayne	31 Oct. 1795	John Mayne	

NINTH LIGHT DRAGOONS—*continued.*

Rank.	1795	Date of Commission.	1796	Date of Commission.	1797	Date of Commission.
Cornet	Lancelot Lauder	30 April 1794	Lancelot Lauder		Lancelot Lauder	
,,	William Martin	,,	Hamilton Bagwell		Hamilton Bagwell	
,,	Hamilton Bagwell	9 May 1794	Thomas Smoke		Brabazon Newcomen	
,,	Thomas Smoke	17 May 1794	W. Deane Staples		George Odlum	
,,	John Mayne	30 May 1794	Brabazon Newcomen		J. L. Fleming	
,,	—— Ellison	31 May 1794	George Odlum	6 Mar. 1795	David West	29 Feb. 1796
,,	Wm. Deane Staples	5 July 1794	George Lynch	3 July 1795	John Wood	,,
,,	Brabazon Newcomen	20 July 1794	John L. Fleming	31 Oct. 1795	Richard Love	30 June 1796
Chaplain	Marcus M'Causland		Marcus M'Causland		Marcus M'Causland	
Adjutant	Richard Johnston		Richard Johnston		Richard Johnston	
Surgeon	William Wallace		William Wallace		William Wallace	
Assistant-Surgeon						
Vet. Surg.						

Rank.	1798	Date of Commission.	1799	Date of Commission.	1800	Date of Commission.
Colonel	Flower Mocher		Flower Mocher		Flower Mocher	
Lieut.-Col.	B. Trecothick Henniker		B. Trecothick Henniker		B. Trecothick Henniker	
,,	Thomas Mahon	1 Jan. 1797	Thomas Mahon		Thomas Mahon	
Major	Humphrey Denis	,,	Humphrey Denis		Humphrey Denis	
,,	Edwd. Witherington	,,	Edwd. Witherington		Edward Witherington	
,,	..		..		Thomas Goldie (Army, 17 Feb. 1798)	20 May 1799
Captain	Richard Donaldson		Richard Donaldson		Richard Donaldson	
,,	John Beevor		John Beevor		John Beevor	
,,	The Hon. William J. Erskine		The Hon. William J. Erskine		Richard Johnston	
					William Needham	
,,	William Martin	1 Dec. 1796	Richard Johnston	28 Feb. 1795	William Williams	19 Jan. 1799
,,	..		William Needham (Army, 12 June 1794)	1 Jan. 1798	Charles Orde	27 April 1799
Capt. and Lieut.	Richard Johnston		Alfred F. Armstrong	29 Dec. 1798	Alfred F. Armstrong	
Lieutenant	Alfred F. Armstrong		John West Wilkins		John West Wilkins	
,,	John West Wilkins		Henry Loftus		Henry Loftus	
,,	Henry Loftus		Robert John Love		Robert John Love	
,,	Robert John Love		Chris. C. Patrickson		Chris. C. Patrickson	
,,	Chris. C. Patrickson		James Lewis Higgins		James Lewis Higgins	
,,	James Lewis Higgins		John Mayne		John Mayne	
,,	John Mayne		Hamilton Bagwell		Hamilton Bagwell	
,,	Hamilton Bagwell	1 Jan. 1797	David West	,,	David West	
Cornet	Launcelot Lauder		Launcelot Lauder		Launcelot Lauder	
,,	Brabazon Newcomen		John L. Fleming		John L. Fleming	
,,	John L. Fleming		John Wood		John Wood	
,,	David West		Richard Love		Richard Love	
,,	John Wood		Henry Witherington		R. Maxwell	
,,	Richard Love		R. Maxwell		Henry Westropp	
,,	Henry Witherington	,,	Henry Westropp	4 Aug. 1798	Henry Dawson	1 April 1799
,,	William Williams	,,	..		William Mathew	15 April 1799
,,	Richard Maxwell	1 June 1797			Samuel Bayly (Army, 25th Oct. 1798)	6 May 1799
Paymaster	..		William Martin	19 May 1798	William Martin	
Adjutant	Richard Johnston		Thomas Butterfield	27 Oct. 1798	Thomas Butterfield	
Surgeon	William Wallace		William Wallace		William Wallace	
Assistant-Surgeon	James Donaghoe	1 May 1797	James Donaghoe		James Donaghoe	
,,	Benjamin Maxwell	,,	Benjamin Maxwell			
Vet. Surg.	..		Samuel Ferguson	4 Oct. 1797	Samuel Ferguson	

NINTH LIGHT DRAGOONS—*continued.*

Rank.	1801	Date of Commission.	1802	Date of Commission.	1803	Date of Commission.
Colonel	Flower Mocher		Sir J. St. Clair Erskine, Bt. (M.-Gen. 1 Jan. 1798)	1 Aug. 1801	Sir J. St. Clair Erskine, Bt.	
Lieut.-Col.	B. Trecothick Heniker		B. Trecothick Heniker		B. Trecothick Heniker (Maj.-Gen. 29 Apr. 1802)	
,,	Thomas Mahon		Thomas Mahon		Thomas Mahon	
Major	Edwd. Witherington		Edwd. Witherington		Edwd. Witherington (Lt.-Col. 29 April 1802)	
,,	Thomas Goldie		Thomas Goldie		Thomas Goldie	
,,	William Needham	14 July 1800	William Needham			
Captain	Richard Donaldson (Lt.-Col., 1 Jan. 1800)		Richard Donaldson		Richard Donaldson	
,,	John Beevor		John Beevor		John Beevor (Maj., 29 Apr. 1802)	
,,	Richard Johnston		Richard Johnston		Richard Johnston	
,,	Charles Orde		Charles Orde		Charles Orde	
,,	Edward Darvall	9 Nov. 1799	Edward Darvall		Edward Darvall	
,,	Robert Aldridge (Army, 9 Nov. 1799)	1 March 1800	Robert Aldridge			
,,	C. Grant	29 Sept. 1800	Robert Ellice	4 May 1801		
Capt. and Lieut.	Alfred F. Armstrong		Alfred F. Armstrong		Alfred F. Armstrong	
Lieutenant	Hamilton Bagwell		Hamilton Bagwell		Hamilton Bagwell	
,,	David West		David West		David West	
,,	John Wood	27 Nov. 1799	Richard Maxwell		Richard Maxwell	
,,	J. L. Fleming	21 Dec. 1799	Richard Gorham		Richard Gorham	
,,	Richard Maxwell	8 Feb. 1800	Henry Westropp		Henry Westropp	
,,	Robert Ellice (Army, 5 Jan. 1799)	15 Feb.	Kingsmill Pennefather	1801	Samuel Bayly	
,,	Richard Gorham	1 March	Samuel Bayly	1 Jan.	John F. Cornwall	
,,	Henry Westropp	,,	John F. Cornwall	7 Jan.		
,,	Kingsmill Pennefather	,,	Alex. Spiers Crawford (Army, 17 Jan. 1799)	4 May		
,,	Richard Love	12 May	Thomas Butterfield	1 June		
Cornet	Henry Dawson		Henry Dawson		John Baldwin	
,,	Samuel Bayly		Thomas Cosby			
,,	John F. Cornwall	1 Aug. 1799	John Baldwin		Benjamin Chapman	
,,	Thomas Butterfield	27 Nov. 1799	Benjamin Chapman	1 July 1800	Thomas Lalor	
,,	Thomas Cosby	18 Jan. 1800	Thomas Lalor		Benjamin Jones	
,,	John Baldwin	12 May 1800	Benjamin Jones	11 Oct. 1800	William Tennison	
,,	Thomas Lalor	26 Aug. 1800	William Tennison	21 Feb. 1801	Samuel White	
,,	..		Samuel White	4 March 1801	Patrick Savage	
,,	..		Patrick Savage	4 May 1801	Richard Rathbone	12 Nov. 1801.
Paymaster	John Mayne	28 Nov. 1799	John Mayne		John Mayne	
Adjutant	Thomas Butterfield		Thomas Butterfield			
Surgeon	William Wallace		William Wallace		William Wallace	
Assistant-Surgeon	James Donaghoe		James Donaghoe		James Donaghoe	
Veterinary Surgeon	Samuel Ferguson		Samuel Ferguson		George Landers	3 July 1802

NINTH LIGHT DRAGOONS—*continued.*

Rank.	1804	Date of Commission.	1805	Date of Commission.	1806	Date of Commission.
Colonel	Sir J. St. Clair Erskine, Bt.		Sir J. St. Clair Erskine, Bt.		James, Earl of Rosslyn (Lieut.-Gen. 1 Jan. 1805	
Lieut.-Col.	B. Trecothick Heniker		B. Trecothick Heniker		B. Trecothick Heniker	
,,	Thomas Mahon		Thomas Mahon		Thomas Mahon (Col. 30 Oct. 1805	
Major	Edward Witherington		Edward Witherington		Edward Witherington	
,,	Thos. Goldie (Lieut.-Col. 25 Sept. 1803)		Thomas Goldie		Thomas Goldie	
Captain	Richard Donaldson		Charles Orde		Charles Orde	
,,	Alfred F. Armstrong	29 Dec. 1798	Edward Darvall		Edward Darvall	
,,	Charles Orde		Samuel Bayly		Robert Ellice	
,,	Edward Darvall		Robert Ellice		Hamilton Bagwell	
,,	Samuel Bayly	12 Mar. 1803	Hamilton Bagwell		Wm. John Sampson	
,,	Robert Ellice (Army, 4 May 1801)	25 May 1803	Wm. John Sampson		Frederick G. Carmichael	
,,	Hamilton Bagwell	25 June 1803	Frederick G. Carmichael	1 Feb. 1804	Matthew G. Blake	
,,	Wm. John Sampson	23 July 1803	Matthew G. Blake	3 Feb. 1804	Richard Maxwell	29 Dec. 1804
,,	..		..		Richard Gorham	14 Feb. 1805
,,	..		..		Samuel White	28 Nov. 1805
Lieutenant	Richard Maxwell		Richard Maxwell		Henry Westropp	
,,	Richard Gorham		Richard Gorham		John F. Cornwall	
,,	Henry Westropp		Henry Westropp		John Baldwin	
,,	John F. Cornwall		John F. Cornwall		William Ebhart	
,,	John Baldwin	12 Mar. 1803	John Baldwin		Wm. Lewis Herries	
	John Philip Perry (Lt. 8 Nov. 1798)	16 April 1803	William Ebhart		Patrick Savage	14 Feb. 1805
,,	Wm. Ebhart (Army, 9 Dec. 1795)	28 July 1803	William B. Tennison	3 Mar. 1804	Richard Rathbone	3 April 1805
,,	Frederick G. Carmichael (Army, 11 Sept. 1800)	1 Oct. 1803	Wm. Lewis Herries (Army, 17 Mar. 1803)	14 April 1804	Wm. Gurley (Army, 24 Nov. 1802)	4 April 1805
,,	..		..		John Hayes	9 May 1805
,,	..		..		Lewis Mackenzie	23 May 1805
,,	..		..		John Clerke (Army, 5 Aug. 1804)	30 May 1805
,,	..		..		Frederick Ebhart	29 Aug. 1805
,,	..		..		Benjamin Handley	7 Nov. 1805
Cornet	Thomas Lalor		Samuel White		Henry Thos. Parker	
,,	Benjamin Jones		Patrick Savage		—— Hutcheon	
,,	William Tennison		Richard Rathbone		John Hames (Army, 17 Mar. 1804	9 May 1805
,,	Samuel White		Henry Thos. Parker		James Maxwell Wallace	14 Aug. 1805
,,	Patrick Savage		John Waring		Thomas Horsbrugh	15 Aug. 1805
,	Richard Rathbone		John Hayes (Army, 25 June 1804)	7 Sept. 1804		
,,	George Ellice	24 July 1802	Lewis Mackenzie	8 Sept. 1804		
,,	Henry Thos. Parker (Army, 10 Oct. 1801)	25 June 1803	—— Hutcheon	26 Oct. 1804		
,,	John Waring (Army, 3 pril 1801)	3 Sept. 1803				
Paymaster	Benjamin Chapman	27 Aug. 1803	Benjamin Chapman		Benjamin Chapman	
Adjutant	John Philip Perry	16 April 1803	—— Hutcheon (Cornet, 20 Oct. 1804)	26 Oct. 1804	—— Hutcheon	
Surgeon	William Wallace		—— Humphries	21 April 1804	Charles Jordan	21 Mar. 1805
Assistant-Surgeon	James Donaghoe		James Donaghoe		James Donaghoe	
,,	..		T. C. Davies	26 Oct. 1804	T. C. Davies	
Veterinary Surgeon	George Landers		George Landers		George Landers	

AGENTS—Messrs. Greenwood and Cox, Craigs Court.

NINTH LIGHT DRAGOONS—*continued.*

Rank.	1807	Date of Commission.	1808	Date of Commission.	1809	Date of Commission.
Colonel	James, Earl of Rosslyn		James, Earl of Rosslyn		James, Earl of Rosslyn	
Lieut.-Col.	B. Trecothick Heniker		B. Trecothick Heniker		B. Trecothick Heniker (Lieut.-Gen. 25 April 1808)	
,,	Thomas Mahon		Thomas Mahon		Thomas Mahon	
Major	Edw. Witherington		Edw. Witherington		Charles Ward Orde	
,,	Charles Ward Orde	17 July 1806	Charles Ward Orde		Robert Ellice	12 May 1808
Captain	Edward Darvall		Robert Ellice		Frederick G. Carmichael	
,,	Robert Ellice		Frederick G. Carmichael		John Baldwin	
,,	Frederick G. Carmichael		Samuel White		George Gore	
,,	Samuel White		George Pigot		Patrick Savage	
,,	George Pigot (Major, 29 April 1802)	3 April 1806	John Baldwin		Lewis Mackenzie	
,,	John Baldwin	12 June 1806	Richard Rathbone		Benjamin Handley	12 May 1808
,,	Richard Rathbone	17 July 1806	Horton Coote Briscoe		Jas. Delancy (Army, 25 Feb. 1808)	19 May 1808
,,	Horton Coote Briscoe (Army, 6 Aug. 1803)	18 Sept. 1806	George Gore		Benjamin Keene (Army, 3 Oct. 1805)	30 June 1808
,,	Charles Dick (Army, 28 Aug. 1806)	25 Sept. 1806	Patrick Savage	12 Mar. 1807	Arthur Gore	14 July 1808
,,	George Gore (Army, 14 Feb. 1805)	25 Sept. 1806	Lewis Mackenzie	25 Sept. 1807	John Godfrey Peters	22 Dec. 1808
Lieutenant	Henry Westropp		Henry Westropp		Henry Westropp	
,,	John F. Cornwall		John F. Cornwall		Wm. Lewis Herries	
,,	Wm. Lewis Herries		William Ebhart		John Clerke	
,,	Patrick Savage		Wm. Lewis Herries		Henry C. Wilson	
,,	Lewis Mackenzie		John Clerke		Henry Thos. Parker	
,,	John Clerke		Benjamin Handley		Joseph Rolfe	
,,	Frederick Ebhart		Henry C. Wilson		Wilkinson Peacock	4 Feb. 1808
,,	Benjamin Handley		Arthur Gore		Denis Mahon	5 April 1808
,,	Henry C. Wilson (Army, 22 May 1804)	2 Jan. 1806	Thomas Horsbrugh		John Knott (Adjut.)	6 April 1808
,,	Arthur Gore (Army, 10 Oct. 1805)	29 May 1806	Henry Thos. Parker		Charles Lucas	7 April 1808
,,	Thomas Horsbrugh	25 July 1806	Joseph Rolfe		Bartholomew B. Thomas	26 May 1808
,,	Thomas M'Dermott (Army, 25 Feb. 1805)	7 Aug. 1806	Thomas Forster		Hon. Francis Bernard	23 June 1808
,,	Henry Thos. Parker		John M'Queen	12 Mar. 1807	Francis Warren	18 Aug. 1808
,,	Joseph Rolfe (Army, 14 Sept. 1804)	9 Oct. 1806	Hon. W. Villiers Mansell (Army, 7 Aug. 1806)	21 May 1807		
,,	Thos. Forster (Army, 4 May 1800)	16 Oct. 1806				
Cornet	John M'Queen	18 June 1806	Wilkinson Peacock		J. Henry Jackson	
,,	Wilkinson Peacock	19 June 1806	Denis Mahon		John Louis Goldsmid	10 Nov. 1808
,,	Denis Mahon	11 Sept. 1806	John Knott (Adjutant).	30 Oct. 1806		
,,	..		Charles Lucas	19 Feb. 1807		
,,	..		J. Henry Jackson (Army, 6 March 1805)	26 Feb. 1807		
Paymaster	Benjamin Chapman		Benjamin Chapman		Benjamin Chapman	
Adjutant	..		John Knott, (Cornet, 30 Oct. 1806)	30 Oct. 1806	John Knott	
Surgeon	Charles Jordan		Charles Jordan		Willam Adams	5 May 1808
Assistant-Surgeon	Richard Swift	25 Sept. 1806	Richard Swift		Richard Swift	
			—— Luby	23 Oct. 1806	—— Luby	
Veterinary Surgeon	George Landers		..		Joseph Gain	24 Mar. 1808

NINTH LIGHT DRAGOONS—*continued.*

Rank	1810	Date of Commission.	1811	Date of Commission.	1812	Date of Commissi
Colonel	James, Earl of Rosslyn		James, Earl of Rosslyn		James, Earl of Rosslyn	
Lieut.-Col.	B. Trecothick Heniker		B. Trecothick Heniker		B. Trecothick Heniker	
,,	The Hon. Thomas Mahon		The Hon. Thomas Mahon		The Hon. Thomas Mahon (Maj.-Gen. 4 June 1811)	
,,	..		William Chabot	24 Jan. 1811	William Chabot	
Major	Charles Ward Orde		Charles Morland (Army, 20 July 1809)	8 Feb. 1810	Charles Morland	
,,	William Chabot (Lt.-Col. 12 Feb. 1807)	1809 16 Mar.				
,,	..		Redmond Morres	24 Jan. 1811	Redmond Morres	
Captain	Frederick G. Carmichael		George Gore		George Gore	
,,	John Baldwin		Patrick Savage		Patrick Savage	
,,	George Gore		Lewis Mackenzie		Lewis Mackenzie	
,,	Patrick Savage		Benjamin Handley		Benjamin Handley	
,,	Lewis Mackenzie		James Delancey		James Delancey	
,,	Benjamin Handley		Benjamin Keene		Benjamin Keene	
,,	James Delancey		John Godfrey Peters		John Godfrey Peters	
,,	Benjamin Keene		Edward Trant Bontein (Army, 15 Dec. 1804)	28 Dec. 1809	Wilkinson Peacock	
					James Hurt	
,,	Arthur Gore		Wilkinson Peacock	10 May 1810	Dennis Daly (Army, 13 May 1807)	1811 10 Oct.
,,	John Godfrey Peters		James Hurt (Army, 25 June 1807)	27 Dec. 1810		
Lieutenant	Henry Westropp		John Clerke		John Clerke	
,,	Wm. Lewis Herries		Henry Thos. Parker		Henry Thos. Parker	
,,	John Clerke		Joseph Rolfe		Joseph Rolfe	
,,	Henry C. Wilson		Denis Mahon		John Knott	
,,	Henry Thos. Parker		John Knott		The Hon. Francis Bernard	
,,	Joseph Rolfe		Bartholomew B. Thomas		Francis Warren	
					Thomas Shaw	
,,	Wilkinson Peacock		The Hon. Francis Bernard		Robert Robertson	
					Benjamin Handley	
,,	Denis Mahon		Francis Warren		John B. Kerr	4 June
,,	John Knott		Thomas Shaw		George Boss	5 June
,,	Bartholomew B. Thomas		Robert Robertson	22 Feb. 1810	Knightly Musg. Clay	6 June
,,	Hon. Francis Bernard		Benjamin Handley	10 May 1810	Henry Webster	13 June
,,	..		..		John William Dunn	3 July
,,	Francis Warren		..		James Grattan	4 July
,,	Thomas Shaw	7 Sept.	..		George Isaac Call (Army, 1 July 1804)	15 Aug.
,,	..		..		Thomas Stannus (Army, 1 Oct. 1805)	4 Sept.
,,	..		..		H. J. Richardson	5 Sept.
,,	..		..		William Barras	19 Sept.
Cornet	R. Robinson (Ensign 20 Oct. 1808)	10 May	John B. Kerr		John Minchin	13 June
,,	Benjamin Handley	11 May	George Boss		John Wright, Adjut.	14 June
,,	John B. Kerr	8 June	Knightly Musg. Clay		Sir Chas. Payne, Bt.	25 July
,,	George Boss (Ensign 15 June 1809)	29 June	..		Lord Guernsey[1]	22 Aug.
,,	Knightly Musg. Clay	12 Oct.	..		James Bogle French	29 Aug.
,,	..		..		Philip Tuite Dalton	3 Oct.
Paymaster	Benjamin Chapman		Benjamin Chapman		Benjamin Chapman	
Adjutant	John Knott		John Knott		John Wright (Cornet 14 June 1811)	4 July
Quarter-master	..		Thomas Hely	2 Nov. 1809	Thomas Hely	
Surgeon	George Peach (Army, 15 Aug. 1805)	15 June	George Peach		George Peach	
Assistant-Surgeon	Richard Swift (Army, 22 Oct. 1800)		Richard Swift		Richard Swift	
,,	—— Luby		John M'Andrew	15 Feb. 1810	John M'Andrew	
Vet. Surg.	Joseph Gain		Joseph Gain		Joseph Gain	

[1] Fifth Earl of Aylesford.

NINTH LIGHT DRAGOONS—*continued.*

Rank.	1813	Date of Commission.	1814	Date of Commission.	1815	Date of Commission.
Colonel	James, Earl of Rosslyn		James, Earl of Rosslyn		James, Earl of Rosslyn (General, 4 June 1814)	
Lieut.-Col.	Hon. B. T. Henniker		Hon. Thos. Mahon		Hon. Thos. Mahon	
,,	Hon. Thos. Mahon		Charles Morland (Army, 29 Oct. 1812)	1813 4 March	Charles Morland	
,,	William Chabot		Simon G. Newport	10 June	Simon Geo. Newport	
Major	Redmond Morres		Redmond Morres		Redmond H. Morres (Lt.-Col., 10 June 1813)	
,,	George Gore	1812 7 May	George Gore		George Gore	
,,	Patrick Savage	29 Oct.	Patrick Savage			
Captain	Benjamin Handley		Benjamin Handley		Benjamin Handley	
,,	James Delancey		James Delancey		James Delancey	
,,	Benjamin Keene		Benjamin Keene		Benjamin Keene	
,,	John Godfrey Peters		John Godfrey Peters		John Godfrey Peters	
,,	Wilkinson Peacock		James Hurt		James Hurt	
,,	James Hurt		Dennis Daly		Dennis Daly	
,,	Dennis Daly		John Clerke		Francis Warren	
,,	John Clerke	7 May	John Gurwood		Simeon H. Stuart (Army, 1 Aug. 1811)	1814 12 Nov.
,,	John Gurwood (Army, 6 Feb. 1812)	30 July	Francis Warren			
,,	Francis Warren	5 Nov.	Alexander Campbell (Army, 13 Aug. 1812)	1 July		
Lieutenant	Henry Thos. Parker		Henry Thos. Parker		Henry Thos. Parker	
,,	Joseph Rolfe		Joseph Rolfe		Joseph Rolfe	
,,	John Knott		Robert Robertson		Robert Robertson	
,,	Hon. Francis Bernard		John B. Kerr		John B. Kerr	
,,	Thomas Shaw		George Boss		George Boss	
,,	Robert Robertson		Henry Webster		Henry Webster	
,,	Benjamin Handley		James Grattan		Thomas Stannus	
,,	John B. Kerr		Thomas Stannus		H. J. Richardson	
,,	George Boss		H. J. Richardson		William Barras	
,,	Henry Webster		William Barras		John Minchin	
,,	James Grattan		John Minchin		J. Wright (Adjutant)	
,,	Thomas Stannus		J. Wright (Adjutant)		Sir Chas. Payne, Bt.	
,,	H. J. Richardson		Sir Chas. Payne, Bt.		Lord George Lennox (Army, 14 Jan. 1813)	18 Aug.
,,	William Barras		Heneage, Earl of Aylesford			
,,	John Minchin	4 Feb.	James Bogle French			
,,	J. Wright (Adjutant)	5 Feb.	Lord George Lennox	14 Jan.		
,,	Sir Chas. Payne, Bt.	6 Feb.	John William Bacon (Army, 28 June 1810)	21 Jan.		
,,	Heneage, Earl of Aylesford	20 May	Richard Beasley.	18 Feb.		
,,	James Bogle French	21 May	Philip Tuite Dalton	1 April		
,,	Isaac Gillam (Army, 6 March 1810)	10 Sept.	W. Bloss Armstrong	29 April		
,,	..		Chas. Davers Allen	20 May		
Cornet	Philip Tuite Dalton		Donald M'Duffie	4 Feb.	Benjamin C. Browne	
,,	Lord George Lennox (Army, 24 Oct. 1811)	13 Feb.	Benjamin C. Browne	17 June	John Gurdon	12 May
,,	W. Bloss Armstrong	12 Mar.	..		Geo. Boss W. Stackpoole	3 June
,,	Richard Beasley	25 June	..		Geo. Gordon Smith	29 Dec.
,,	Charles Davis Allen	17 Sept.				
Paymaster	Benjamin Chapman		Benjamin Chapman		Benjamin Chapman	
Adjutant	John Wright (Army, Lieut. 5 Feb. 1812)		John Wright		John Wright	
Qr.-master	Thomas Hely		Thomas Hely		Thomas Hely	
Surgeon	George Peach		George Peach		Jas. O'Connor (Army, 3 Feb. 1803	16 Feb. 1815
Assistant-Surgeon	John M'Andrew		John M'Andrew		John M'Andrew	
,,	Geo. Purdon (Army, 7 May 1807)	22 Oct.	Thos. Lukis (Army, 9 April 1812)	25 Feb.	James Knox (Army, 12 Nov. 1812)	25 Aug. 1814
Veterinary Surgeon	Joseph Gain		—— Korke	11 Mar.	J. Norton	2 Oct. 1814

Blue. Facings buff. Lace silver. AGENTS—Messrs. Greenwood, Cox, and Hammersley.

Blue. Facings crimson. Lace gold.

ENGLISH AGENTS — Messrs. Greenwood, Cox, and Hammersley.

NINTH LANCERS.

Rank.	1816	Date of Commission.	1817	Date of Commission.	1818	Date of Commission.
Colonel	James, Earl of Rosslyn		James, Earl of Rosslyn		James, Earl of Rosslyn	
Lieut.-Col.	Hon. Thomas Mahon		Hon. Thomas Mahon		Hon. Thomas Mahon	
,,	Charles Morland		Charles Morland		Charles Morland	
,,	Simon G. Newport		Simon G. Newport			
Major	R. H. de Montmorency (Lt.-Col. 10 June 1813)	24 Jan. 1811	George Gore		[W.] Thomas Wildman [1]	
,,	George Gore		Thomas Wildman (Army, 18 July 1816)	31 Oct. 1816	James Delancey	4 Dec. 1817
Captain	Benjamin Handley		Benjamin Handley		James Hurt	
,,	James Delancey		James Delancey		Dennis Daly	
,,	Benjamin Keene		John Godfrey Peters		Alexander Campbell	
,,	John Godfrey Peters		James Hurt		[W.] Lord G. Lennox	
,,	James Hurt		Dennis Daly		John B. Kerr	2 Jan. 1817
,,	Dennis Daly		Francis Warren		Sir Charles Payne, Bt.	10 April 1817
,,	Francis Warren		Alexander Campbell		Augustus F. d'Esté (Army, 14 Mar. 1815)	24 April 1817
,,	Alexander Campbell		Lord George Lennox (Army, 9 Nov. 1815)	27 June 1816	H. J. Richardson	4 Dec. 1817
Lieutenant	Henry Thos. Parker		Henry Thos. Parker		George Boss	
,,	Joseph Rolfe		Robert Robertson		William Barras	
,,	Robert Robertson		John B. Kerr		John Minchin	
,,	John B. Kerr		George Boss		John Wright (Adjutant)	
,,	George Boss		Thomas Stannus		George Anfrere	
,,	Henry Webster		H. J. Richardson		Benjamin C. Browne	2 Jan. 1817
,,	Thomas Stannus		William Barras		George Gordon Smith	10 April 1817
,,	H. J. Richardson		John Minchin		Charles Bacon (Army, 20 May 1813)	26 June 1817
,,	William Barras		John Wright (Adjutant)		W. Leader Maberly (Army, 23 Mar. 1815)	3 July 1817
,,	John Minchin		Sir Charles Payne, Bt.		[W.] William War. Huntley (Army, 23 Oct. 1815)	30 Oct. 1817
,,	John Wright (Adjutant)		George Anfrere (Army, 9 Mar. 1815)	5 Sept. 1816	George M'Dowall	4 Dec. 1817
,,	Sir Charles Payne, Bt.					
Cornet	Benjamin C. Browne		Benjamin C. Browne		Montagu H. Grant	
,,	John Gurdon		Geo. Gordon Smith		Robert Norris	
,,	Geo. Boss W. Stackpoole		George M'Dowall	4 April 1816	Julius Greenwood	
,,	Geo. Gordon Smith		Montague Harvey Grant	18 April 1816	Aug. Frederick Ellis	10 April 1817
,,	..		Robert Norris	9 May 1816	Henry Ed. Porter	3 July 1817
,,	..		Julius Greenwood	5 Sept. 1816	John Whyte-Melville	4 Dec. 1817
Paymaster	Benjamin Chapman		Benjamin Chapman		Benjamin Chapman	
Adjutant	John Wright		John Wright		John Wright	
Quarter-master	Thomas Hely		Thomas Hely		Thomas Hely	
Surgeon	James O'Connor		James O'Connor		James O'Connor	
Assistant-Surgeons	John M'Andrew		John M'Andrew		John M'Andrew	
	James Knox		James Knox		James Knox	
Veterinary Surgeon	J. Norton		J. Norton		J. Norton	

[1] The letter "W" indicates that the officer is in possession of the Waterloo Medal. Being present at the actions on the 16th, 17th, or 18th June 1815.

NINTH LANCERS—*continued.*

Rank.	1819	Date of Commission.	1820	Date of Commission.	1821	Date of Commission.
Colonel	James, Earl of Rosslyn		James, Earl of Rosslyn		James, Earl of Rosslyn, G.C.B.	
Lieut.-Col.	Hon. Thomas Mahon		Charles Morland		Charles Morland	
,,	Charles Morland					
Major	[W.] Thos. Wildman		H. F. C. Cavendish		H. F. C. Cavendish	
,,	H. F. C. Cavendish (Army, 2 Apr. 1818)	24 Sept. 1818	James Hurt	23 Dec. 1819	James Hurt	
Captain	James Hurt		Dennis Daly		Dennis Daly	
,,	Dennis Daly		Alexander Campbell		Alexander Campbell	
,,	Alexander Campbell		[W.] Lord George Lennox		[W.] Lord George Lennox	
,,	[W.] Lord George Lennox (25 Oct. 1817)		John B. Kerr		John B. Kerr	
,,	John B. Kerr		Sir Charles Payne, Bt.		Augustus F. d'Esté	
,,	Sir Charles Payne, Bt.		Augustus F. d'Esté		H. J. Richardson	
,,	Augustus F. d'Esté		H. J. Richardson		John Minchin	
,,	H. J. Richardson		John Minchin	,,	Charles Blakiston (Army, 14 Mar. 1816)	15 June 1820
Lieutenant	George Boss		John Wright (Adjutant)		John Wright (Adjutant)	
,,	John Minchin		Benjamin C. Browne		Benjamin C. Browne	
,,	John Wright (Adjutant)		Charles Bacon		Charles Bacon	
,,	George Anfrere		[W.] W. W. Huntley		[W.] W. W. Huntley	
,,	Benjamin C. Browne		George M'Dowall		George M'Dowall	
,,	Charles Bacon		Julius Greenwood		Julius Greenwood	
,,	[W.] W. W. Huntley		Augustus Fred. Ellis		Augustus Fred. Ellis	
,,	George M'Dowall		Henry Ferguson	18 Mar. 1819	Henry Ferguson	
,,	Julius Greenwood	11 June 1818	Henry Mallory (Army, 6 Jan. 1819)	4 Nov. 1819	Henry Mallory	
,,	Augustus Fred. Ellis (Army, 17 Dec. 1818)	31 Dec. 1818	Henry Ed. Porter	23 Dec. 1819	Henry Ed. Porter	
Cornet	Henry Ed. Porter		Henry Knight		Henry Knight	
,,	John Whyte-Melville		Richard Wright		Richard Wright	
,,	Henry Ferguson	18 Mar. 1818	Chas. Tyrwhitt Jones		Chas. Tyrwhitt Jones	
,,	Henry Knight	19 Mar. 1818	Sir Francis Vincent, Bt.		Sir Francis Vincent, Bt.	
,,	Richard Wright (Army, 24 Jan. 1818)	16 April 1818	Lord Geo. Bentinck (Lt. 26 Nov. 1818)	18 Feb. 1819	Lord Geo. Bentinck	
,,	Charles Tyrwhitt Jones	11 June 1818	Jas. A., Lord Loughborough	25 Feb. 1819	William Beresford	
,,	Sir Francis Vincent, Bt.	17 Dec. 1818	William Beresford	23 Dec. 1819	J. Knightley Musgrave	16 Nov. 1820
Paymaster	Benjamin Chapman		[W.] Henry Knight (Army, 2 July 1805)	,,	[W.] Henry Knight	
Adjutant	John Wright		John Wright		John Wright	
Qr.-master	Thomas Hely		Thomas Hely		Thomas Hely	
Surgeon	James O'Connor		Jas. O'Connor, M.D.		Jas. O'Connor, M.D.	
Assistant-Surgeon	Edward Burton (Army, 21 Jan. 1813)	23 July 1818	Edward Burton		Edward Burton	
Veterinary Surgeon	J. Norton		J. Norton		J. Norton	

NINTH LANCERS—*continued.*

Rank.	1822	Date of Commission.	1823	Date of Commission.	1824	Date of Commission.
Colonel	James, Earl of Rosslyn, G.C.B.,		James, Earl of Rosslyn, G.C.B.		James, Earl of Rosslyn, G.C.B.	
Lieut.-Col.	Charles Morland		Charles Morland		Charles Morland	
Major	James Hurt		James Hurt		James Hurt	
,,	Alexander Campbell	1821 9 Aug.	Alexander Campbell		Alexander Campbell	
Captain	Dennis Daly (Major, 19 July 1821)		Dennis Daly		H. J. Richardson	
,,	[W.] Lord George Lennox		[W.] Lord G. Lennox		Charles H. Somerset	
,,	Augustus F. d'Esté		H. J. Richardson		Benjamin C. Browne	
,,	H. J. Richardson		Charles H. Somerset		J. A., Lord Loughborough (Army, 25 March 1823)	1823 22 May
,,	John Minchin		Benjamin C. Brown	1822 18 July	Julius Greenwood	5 June
,,	Charles H. Somerset (Army, 17 June 1819)	15 Nov.	Henry Ferguson	26 Sept.	Henry Mallory	18 Dec.
Lieutenant	John Wright (Adjutant)		John Wright (Adjutant)		John Wright (Adjutant)	
,,	Benjamin C. Browne		Julius Greenwood		Henry Edward Porter	
,,	George M'Dowall		Henry Mallory		Thomas Montgomery	
,,	Julius Greenwood		Henry Edward Porter		Richard Wright	
,,	Henry Ferguson		Henry Knight		William Beresford	
,,	Henry Mallory		Thomas Montgomery (Army, 4 Oct. 1821)	2 May	J. Knightley Musgrave	5 June
,,	Henry Edward Porter		Richard Wright	18 July	The Hon. George Vaughan	17 July
,,	Henry Knight	13 Sept.	William Beresford	26 Sept.	Robert Markham	18 Dec.
Cornet	Richard Wright		J. Knightley Musgrave		James Alex. Fullerton	
,,	William Beresford		The Hon. George Vaughan		Percy B. Williams	
,,	J. Knightley Musgrave		The Hon. Frederick Lascelles		John Rind	
,,	The Hon. George Vaughan	10 May	James Alex. Fullerton	1 Aug.	Rich. F. Shawe	5 June
,,	The Hon. Frederick Lascelles	12 July	Percy B. Williams	26 Sept.	George Knox (Ensign, 15 May 1823)	17 July
,,	..		John Rind (Ensign, 29 April 1813; Riding-Master)	25 Oct.	T. John, Viscount Kirkwall	18 Sept.
Paymaster	[W.] Henry Knight		[W.] Henry Knight		[W.] Henry Knight	
Adjutant	John Wright		John Wright		John Wright	
Qr.-master	Thomas Hely		Thomas Hely		Thomas Hely	
Surgeon	Jas. O'Connor, M.D.		Anthony Cæsar Colclough (Army, 14 April 1813)	17 Jan.	Anthony Cæsar Colclough	
Assistant-Surgeon	Edward Burton		Edward Burton		Edward Burton	
Veterinary Surgeon	J. Norton		J. Norton		J. Norton	

NINTH LANCERS—*continued.*

Rank.	1825	Date of Commission.	1826	Date of Commission.	1827	Date of Commission.
Colonel	James, Earl of Rosslyn, G.C.B.		James, Earl of Rosslyn, G.C.B.		James, Earl of Rosslyn, G.C.B.	
Lieut.-Col.	Charles Morland		Charles Morland (Colonel 27 May 1825)		Charles Morland	
Major	James Hurt		Alexander Campbell		Alexander Campbell	
,,	Alexander Campbell		H. J. Richardson	14 July 1825	J. A., Lord Loughborough	12 Dec. 1826
Captain	H. J. Richardson		Charles H. Somerset		Julius Greenwood	
,,	Charles H. Somerset		Benjamin C. Browne		Henry E. Porter	
,,	Benjamin C. Browne		J. A., Lord Loughborough		Richard Wright	1 Aug. 1826
,,	J. A., Lord Loughborough		Julius Greenwood		James A. Fullerton	19 Sept. 1826
,,	Julius Greenwood		Henry Mallory		Henry J. Ramsden (Army, 30 June 1825)	2 Nov. 1826
,,	Henry Mallory		Henry E. Porter	14 July 1825	Percy B. Williams	12 Dec. 1826
Lieutenant	John Wright (Adjutant)		John Wright (Adjutant)		John Wright (Adjutant)	
,,	Henry E. Porter		Richard Wright		John Rind	
,,	Richard Wright		Hon. G. Vaughan		Richard Fleetwood-Shawe	
,,	William Beresford		James A. Fullerton		The Hon. W. E. Fitzmaurice	
,,	J. Knightley Musgrave		Percy B. Williams	26 May 1825	Randall Rumley	
,,	The Hon. George Vaughan		John Rind (Riding-Master)	13 July 1825	George Vesey	1 Aug. 1826
,,	Robert Markham		Richard Fleetwood-Shawe	14 July 1825	Arthur C. Williams	29 Aug. 1826
,,	James A. Fullerton	30 Dec. 1824	The Hon. W. E. Fitzmaurice	24 Sept. 1825	Nathaniel Weekes	19 Sept. 1826
,,	..		..		Edward Spencer Trower	12 Dec. 1826
Cornet	Percy B. Williams		Randall Rumley	8 Oct. 1825	Frederick Willis	
,,	John Rind		George Vesey	15 June 1825	Alexander, Viscount Fincastle	1 Aug. 1826
,,	Richard Fleetwood-Shawe		Arthur C. Williams	16 June 1825	James Hope Grant	29 Aug. 1826
,,	George Knox		Nathaniel Weekes	24 Sept. 1825	John Menzies	19 Sept. 1826
,,	The Hon. W. E. Fitzmaurice	18 Nov. 1824	Edward Spencer Trower	29 Sept. 1825	John Meiklam	12 Dec. 1826
,,	Randall Rumley	30 Dec. 1824	Frederick Willis	3 Nov. 1825		
Paymaster	[W.] Henry Knight		[W.] Henry Knight		[W.] Henry Knight	
Adjutant	John Wright		John Wright		John Wright	
Quarter-master	Thomas Hely		T. Wright	11 Aug. 1825	T. Wright	
Surgeon	Anthony Cæsar Colclough		Anthony Cæsar Colclough		Anthony Cæsar Colclough	
Assistant-Surgeon	Edward Burton		Edward Burton		W. J. Shiell, M.D. (Army, 26 Sept. 1822)	15 June 1826
Veterinary Surgeon	J. Norton		Thomas Harrison	14 Apr. 1825	Thomas Harrison	

NINTH LANCERS—*continued.*

Rank.	1828	Date of Commission.	1829	Date of Commission.
Colonel	James, Earl of Rosslyn, G.C.B.		James, Earl of Rosslyn, G.C.B.	
Lieut.-Col.	Charles Morland		James Alex., Lord Loughborough (Army, 31 Dec. 1827)	6 March 1828
Major	Alexander Campbell		Alexander Campbell	
,,	Julius Greenwood	31 Dec. 1827	Julius Greenwood	
Captain	Henry Edw. Porter		Henry Edw. Porter	
,,	James A. Fullerton		James A. Fullerton	
,,	Henry J. Ramsden		Henry J. Ramsden	
,,	Percy B. Williams		Percy B. Williams	
,,	Richard F. Shawe	27 April 1827	Richard F. Shawe	
,,	George Callaghan	31 Dec. 1827	The Hon. W. E. Fitz-Maurice	26 Feb. 1828
Lieutenant	John Rind		John Rind	
,,	The Hon. W. E. Fitz-Maurice		George Vesey	
,,	George Vesey		Arthur C. Williams	
,,	Arthur C. Williams		Nathaniel Weekes	
,,	Nathaniel Weekes		Edward S. Trower	
,,	Edw. Spencer Trower		Lewis Upton	
,,	Lewis Upton (Army, 11 July 1826)	22 Mar. 1827	Frederick Willis	
,,	Frederick Willis	27 April 1827	James Hope Grant	,,
,,	Alexander, Viscount Fincastle	14 Aug. 1827	E. St. Vincent Digby (Army, 27 Apr. 1827)	3 July 1828
Cornet	James Hope Grant		John Meiklam	
,,	John Menzies		R. Cooke (Adjutant)	
,,	John Meiklam		John Eden Spalding	
,,	Robert Cooke (Adjutant)	27 April 1827	William Hankey	3 Jan. 1828
,,	John Eden Spalding (Army, 12 Apr. 1827)	14 Aug. 1827	H. Willoughby Legard	26 Feb. 1828
Paymaster	[W.] Henry Knight		[W.] Henry Knight	
Adjutant	Robert Cooke (Cornet 27 April 1827)	27 April 1827	Robert Cooke	
Qr.-master	T. Wright		T. Wright	
Surgeon	A. C. Colclough		A. C. Colclough	
Assistant-Surgeon	W. Jas. Shiell, M.D.		William I. Breslin (Army, 8 Mar. 1827)	27 Nov. 1828
Veterinary Surgeon	Thomas Harrison		Thomas Harrison	

NINTH (QUEEN'S ROYAL) LANCERS.

Rank.	1830	Date of Commission.	1831	Date of Commission.	1832	Date of Commission.
Colonel	James, Earl of Rosslyn, G.C.B.		James, Earl of Rosslyn, G.C.B.		James, Earl of Rosslyn, G.C.B.	
Lieut.-Col.	James Alex., Lord Loughborough		James Alex., Lord Loughborough		James Alex., Lord Loughborough	
Major	Alexander Campbell		Alexander Campbell (Lt.-Col., 16 July 1830)		Alex. Campbell, K.H.[1]	
,,	Julius Greenwood		Julius Greenwood		Henry Edw. Porter	4 Oct. 1831
Captain	Henry Edw. Porter		Henry Edw. Porter		James Alex. Fullerton	
,,	James Alex. Fullerton		James Alex. Fullerton		Percy B. Williams	
,,	Henry J. Ramsden		Henry J. Ramsden		George Vesey	15 Mar. 1831
,,	Percy B. Williams		Percy B. Williams		Arthur C. Williams	5 July 1831
,,	Richard F. Shawe		Richard F. Shawe		Nathaniel Weekes	4 Oct. 1831
,,	The Hon. W. E. Fitz-Maurice		The Hon. W. E. Fitz-Maurice		John Carnegie (Army, 8 June 1830)	6 Jan. 1832
Lieutenant	John Rind		John Rind		Edward S. Trower	
,,	George Vesey		George Vesey		Lewis Upton	
,,	Arthur C. Williams		Arthur C. Williams		Frederick Willis	
,,	Nathaniel Weekes		Nathaniel Weekes		James Hope Grant	
,,	Edward S. Trower		Edward S. Trower		John Meiklam	15 Mar. 1831
,,	Lewis Upton		Lewis Upton		Robert Cooke (Adjutant)	5 April 1831
,,	Frederick Willis		Frederick Willis		William Hankey	10 May 1831
,,	James Hope Grant		James Hope Grant		H. Willoughby Legard	5 July 1831
,,	Edward St. Vincent Digby		Edward St. Vincent Digby		David L. Campbell	4 Oct. 1831
,,	..		..		Robert E. Fullerton	20 Jan. 1832
Cornet	John Meiklam		John Meiklam		James N. Macartney	10 May 1831
,,	Robert Cooke (Adjutant)		Robert Cooke (Adjutant)		Frederick Gerard	6 July 1831
	John Eden Spalding		John Eden Spalding		Archibald Little	4 Oct. 1831
,,	William Hankey		William Hankey		Hen. Armytage Whatman	20 Jan. 1832
,,	H. W. Legard		H. W. Legard			
Paymaster	[W.] Henry Knight		[W.] Henry Knight		[W.] Henry Knight	
Adjutant	Robert Cooke		Robert Cooke		Robert Cooke	
Quarter-master	T. Wright		T. Wright		T. Wright	
Surgeon	Anth. C. Colclough		George Rich. Melin (Forces, 1 Sept. 1825)	19 Nov. 1830	George Rich. Melin	
Assistant-Surgeon	William I. Breslin		William I. Breslin		William I. Breslin	
Vet. Surg.	James Robertson	24 Sept. 1829	James Robertson		James Robertson	

[1] K.H. indicates a Knight of the Hanoverian-Guelphic Order.

Scarlet. Facings blue. Lace gold.
AGENTS—Messrs. Greenwood, Cox, and Co.

NINTH (QUEEN'S ROYAL) LANCERS—*continued.*

Rank.	1833	Date of Commission.	1834	Date of Commission.	1835	Date of Commission.
Colonel	James, Earl of Rosslyn, G.C.B.		James, Earl of Rosslyn, G.C.B.		James, Earl of Rosslyn, G.C.B.	
Lieut.-Col.	Jas. Alex., Lord Loughborough		Jas. Alex., Lord Loughborough		Jas. Alex., Lord Loughborough	
Major	Alexander Campbell, K.H.		Alexander Campbell, K.H.		Alexander Campbell, K.H.	
,,	Henry Ed. Porter	4 Oct. 1831				
Captain	Jas. Alex. Fullerton		James Alex. Fullerton		Jas. Alex. Fullerton	
,,	Percy B. Williams		Percy B. Williams		Percy B. Williams	
,,	Arthur C. Williams		Arthur C. Williams		Arthur C. Williams	
,,	Nathaniel Weekes		Nathaniel Weekes		Nathaniel Weekes	
,,	John Carnegie		John Carnegie		Edward S. Trower	
,,	Edward Spencer Trower	4 May 1832	Edward Spencer Trower		Frederick Willis	25 April 1834
Lieutenant	Frederick Willis		Frederick Willis		Jas. Hope Grant	
,,	James Hope Grant		Jas. Hope Grant		R. Cooke (Adjutant)	
,,	John Meiklam		R. Cooke (Adjutant)		William Hankey	
,,	Robert Cooke		William Hankey		H. Willoughby Legard	
,,	William Hankey		H. Willoughby Legard		Archibald Little	
,,	H. Willoughby Legard		Archibald Little		H. Armytage Whatman	
,,	David Lyon Campbell		H. Armytage Whatman	5 April 1833	Thomas P. Whalley	
,,	Frederick Gerard	,,	Thomas P. Whalley	12 April 1832	James Clerk	
,,	Bingley Broadhead (Army, 20 Sept. 1826)	18 May 1832	Sir Joseph Hawley, Bt.	28 June 1832	James N. Macartney	11 April 1834
,,	Archibald Little	31 Aug. 1832	James Clerk (Army, 11 June 1833)	24 Dec. 1832	Thomas Porter	25 April 1834
Cornet	Jas. Nixon Macartney		James N. Macartney (Riding-Master)		Charles H. Ibbetson	
,,	H. Armytage Whatman		Thomas Porter	5 April 1833	James Johnston	
,,	Thomas Palmer Whalley	4 May 1832	Charles H. Ibbetson	12 April 1833	Andrew Spottiswoode (Ensign, 10 May 1833)	,,
,,	Sir Joseph Hawley, Bt.	31 Aug. 1832	James Johnston	12 July 1833		
Paymaster	[W.] Henry Knight		[W.] Henry Knight		[W.] Henry Knight	
Adjutant	Robert Cooke		Robert Cooke		Robert Cooke	
Qr.-master	T. Wright		T. Wright		T. Wright	
Surgeon	George Rich. Melin		George Rich. Melin		George Rich. Melin	
Assistant-Surgeon	William Irwin Breslin		William Irwin Breslin		William Irwin Breslin	
Vet. Surg.	James Robertson		James Robertson		James Robertson	

Scarlet. Facings blue. Do. AGENTS—Messrs. Cox and Co.

Rank.	1836	Date of Commission.	1837	Date of Commission.	1838	Date of Commission.
Colonel	James, Earl of Rosslyn, G.C.B.		James, Earl of Rosslyn, G.C.B. (died 18 Jan. 1837)		Samuel Need (Lieut.-Gen., 22 July 1830)	30 Jan 1837
Lieut.-Col.	James Alex., Lord Loughborough		James Alex., Earl of Rosslyn		James Alex., Earl of Rosslyn	
Major	Alexander Campbell, K.H.		Alexander Campbell, K.H.		Alexander Campbell, K.H.	
Captain	James A. Fullerton		James A. Fullerton		James A. Fullerton	
,,	Percy B. Williams		Percy B. Williams		Percy B. Williams	
,,	Arthur C. Williams		Arthur C. Williams		Arthur C. Williams	
,,	Nathaniel Weekes		Nathaniel Weekes		J. Hope-Grant	
,,	Edward S. Trower		J. Hope Grant		William Hankey	
,,	J. Hope Grant	29 May 1835	William Hankey	12 Feb. 1836	Archibald Little	24 Feb. 1837
Lieutenant	R. Cooke (Adjutant)		R. Cooke (Adjutant)		R. Cooke (Adjutant)	
,,	William Hankey		Archibald Little		Thomas P. Whalley	

NINTH (QUEEN'S ROYAL) LANCERS—*continued.*

Rank.	1836	Date of Commission.	1837	Date of Commission.	1838	Date of Commission.
Lieutenant	H. Willoughby Legard		Thomas P. Whalley		James N. Macartney (Riding-Master)	
,,	Archibald Little		James Clerk		James Johnston	
,,	H. Armytage Whatman		James N. Macartney (Riding-Master)		Andrew Spottiswoode	
,,	Thomas P. Whalley		Thomas Porter		William S. A. Ellis	
,,	James Clerk		James Johnston	1836 12 Feb.	John A. Thomson	1837 24 Feb.
,,	James N. Macartney		Andrew Spottiswoode	29 April	J. W. Gooch Spicer	5 May
,,	Thomas Porter		William Sam. A. Ellis (Army, 6 Dec. 1833)	23 Dec.	Frederick James Isacke	17 Nov.
,,	Charles H. Ibbetson	1835 29 May				
Cornet	James Johnston		John A. Thomson	12 Feb.	F. D. Willoughby	
,,	Andrew Spottiswoode		J. W. Gooch Spicer	29 April	Kingsmill M. Power	24 Feb.
,,	Robert Meiklam	,,	Frederick J. Isacke	2 Sept.	Philip H. Clarke	5 May
,,	..		F. D. Willoughby	16 Sept.	John Sutherland	17 Nov.
Paymaster	[W.] Henry Knight		[W.] Henry Knight		[W.] Henry Knight	
Adjutant	Robert Cooke		Robert Cooke		Robert Cooke	
Qr.-master	Thomas Wright		Thomas Wright	11 Aug. 1825	Thomas Wright	
Surgeon	George Rich. Melin		Anthony C. Colclough		Anthony C. Colclough	
Asst.-Surg.	William Irwin Breslin		William Irwin Breslin		William Irwin Breslin	
Vet. Surg.	James Robertson		James Robertson		James Robertson	

Rank.	1839	Date of Commission.	1840	Date of Commission.	1841	Date of Commission.
Colonel	Samuel Need		[W.] ✠ Jas. W. Sleigh (Maj.-Gen. 22 July 1830)[1]	24 Aug. 1839	[W.] ✠ Jas. Wallace Sleigh	
Lieut.-Col.	Jas. Alexander, Earl of Rosslyn		Jas. Alexander, Earl of Rosslyn		Jas. Alexander, Earl of Rosslyn	
Major	..		Alex. Campbell, K.H.		Alex. Campbell, K.H.	
Captain	James A. Fullerton		James A. Fullerton		James A. Fullerton	
,,	Percy B. Williams		Arthur C. Williams		Arthur C. Williams	
,,	Arthur C. Williams		James Hope Grant		James Hope Grant	
,,	James Hope Grant		Archibald Little		Archibald Little	
,,	Archibald Little		T. Palmer Whalley		Andrew Spottiswoode	
,,	T. Palmer Whalley	17 Aug. 1838	Andrew Spottiswoode	11 Oct. 1839	J. Anstruther Thomson	15 Jan. 1841
Lieutenant	R. Cooke (Adjutant)		R. Cooke (Adjutant)		R. Cooke (Adjutant)	
,,	James N. Macartney		James N. Macartney (Riding-Master)		James N. Macartney (Riding-Master)	
,,	Andrew Spottiswoode		W. Samuel Augustine Ellis		J. W. Gooch Spicer	
,,	W. Samuel Augustine Ellis		J. Anstruther Thomson		Frederick J. Isacke	
,,	J. Anstruther Thomson		J. W. Gooch Spicer		F. D. Willoughby	
,,	J. W. Gooch Spicer		Frederick J. Isacke		Kingsmill M. Power	
,,	Frederick J. Isacke		F. D. Willoughby		Philip H. Clarke	
,,	F. D. Willoughby	9 Feb. 1838	Kingsmill Manley Power		John Sutherland	31 Jan. 1840
,,	Kingsmill Manley Power	17 Aug. 1838	Philip H. Clarke	,,	Gilbert T. Nicholson	15 Jan. 1841
Cornet	Philip H. Clarke		John Sutherland		John Edw. Madocks	
,,	John Sutherland		Gilbert T. Nicholson		Eustace Arkwright	31 Jan. 1840
,,	Gilbert T. Nicholson	9 Feb. 1838	John Edw. Madocks		Arthur H. Jenney	10 Apr. 1840
,,	John E. Madocks	17 Aug. 1838	Hon. Brownlow Chas. Bertie	,,		
Paymaster	[W.] Henry Knight		[W.] Henry Knight		[W.] Henry Knight	
Adjutant	Robert Cooke		Robert Cooke		Robert Cooke	
Qr.-master	J. Clark Williamson	,,	John C. Williamson		John C. Williamson	
Surgeon	Anthony C. Colclough		Anthony C. Colclough		C. T. Ingham, M.D. (Army, 25 June 1826)	21 Aug. 1840
Asst.-Surg.	William I. Breslin		William I. Breslin		William I. Breslin	
Vet. Surg.	James Robertson		James Robertson		James Robertson	

[1] ✠ This indicates that the officer is a Companion of the "Order of the Bath."

AGENT—Mr. Collyer.

NINTH (QUEEN'S ROYAL) LANCERS—*continued.*

Rank.	1842	Date of Commission.	1843	Date of Commission.	1844	Date of Commission.
Colonel	[W.] ✠James Wallace Sleigh (Lt.-Gen. 23 Nov. 1841)		[W.] ✠James Wallace Sleigh		[W.] ✠James Wallace Sleigh	
Lieut.-Col.	Alexander Campbell, K.H.	28 Jan. 1842	Alexander Campbell, K.H.		Alexander Campbell, K.H.	
,,	..		✠John Scott, (Army, 31st August 1830)	22 April 1842	✠John Scott	
Major	James Alex. Fullerton (Army, 23 Nov. 1841)	28 Jan. 1842	James Alex. Fullerton		James Alex. Fullerton	
,,	..		✠James Hope Grant	,,	✠James Hope Grant	
Captain	James Hope Grant		Archibald Little		Archibald Little	
,,	Archibald Little		Andrew Spottiswoode		Andrew Spottiswoode	
,,	Andrew Spottiswoode		[W.] Wm. Warburton Huntley		Edward Rudston Read	
,,	William Warburton Huntley (Major, 10 Jan. 1837)	16 July 1841	Edward Rudston Read		Thomas Westropp M'Mahon	
,,	Edward Rudston Read (Army, 27 Aug. 1841)	12 Oct. 1841	Thomas Westropp M'Mahon (Army, 9 June 1838)	,,	Charles Massey Dawson	
,,	Frederick Jas. Isacke	28 Jan. 1842	Lucius Henry French (Army, 28 June 1839)	29 April 1842	John Rose Holden Rose	
,,	..		Charles Massey Dawson (Army, 26 Oct. 1841)	,,	Francis Digby Willoughby	
,,	..		John Rose Holden Rose	,,	James Nixon Macartney	13 Nov. 1842
,,	..		Francis Digby Willoughby	13 May 1842	Kingsmill Manley Power	24 Nov. 1843
Lieutenant	Robert Cooke (Adjutant)		James Nixon Macartney (Riding-Master)		Gilbert Thomas Nicholson	
,,	James Nixon Macartney (Riding-Master		Kingsmill Manley Power		John Cameron Campbell	
,,	Francis Digby Willoughby		Gilbert Thomas Nicholson		Chas. Henry Douglas Donovan	
,,	Kingsmill Manley Power		John Cameron Campbell		The Hon. Charles Powys	
,,	Gilbert Thomas Nicholson		Chas. Henry Douglas Donovan		William Drysdale	
,,	Eustace Arkwright	11 June 1841	Norton Thomas Williams (Army, 14 Jan. 1842)	4 Mar. 1842	Octavius Henry St. George Anson	
,,	John Cameron Campbell (Army, 16 Mar. 1832)	18 June 1841	The Hon. Charles Powys (Army, 1 July 1836)	22 April 1842	William Wellington Waterloo Humbley	
,,	Chas. Henry Douglas Donovan (Army, 3 Aug. 1841)	29 Oct. 1841	William Drysdale (Army, 31 Aug. 1838)	,,	Charles John Colville	
,,	Arthur Henry Jenney	28 Jan. 1842	Octavius Henry St. George Anson (Army, 8 Dec. 1838)	,,	Wm. Richard Newport Campbell	
,,	..		William Wellington Waterloo Humbley (Army, 15 Dec. 1838)	,,	J. K. Wedderburn	
,,	..		Charles John Colville (Army, 21 Feb. 1840)	,,	Philip Kemp	
,,	..		Wm. Rd. Newport Campbell (Army, 15 July 1841)	,,	Edmund John Turner	
,,	..		J. K. Wedderburn, (Army, 10 Apr. 1840)	29 April 1842	James Wm. Robarts	
,,	..		Philip Kemp (Army, 2 Sept. 1836)	6 May 1842	Frederick Courtney Trower	
,,	..		Edmund John Turner (Army, 7 Feb. 1840)	,,	Thos. John Francis	
,,	..		James William Robarts (Army, 31 Dec. 1841)	,,	Wm. Francis Anderton (Riding-Master)	13 Nov. 1842

NINTH (QUEEN'S ROYAL) LANCERS—*continued.*

Rank.	1842	Date of Commission.	1843	Date of Commission.	1844	Date of Commission.
Lieutenant	..		Frederick Courtney Trower	13 May 1842	Robert Stordy Dixon	23 June 1843
,,	..		Thomas John Francis (Army, 1 July 1842)	27 Sept. 1842	Charles Fitzgerald Studdert	24 Nov. 1843
Cornet	Wm. Henry Magan	5 Feb. 1841	William Francis Anderton		Philip Antrobus	
,,	William Francis Anderton	18 May 1841	John Hallowell Carew		Alexander Hawtrey	
,,	John Hallowell Carew	11 June 1841	Robert Stordy Dixon		William Coesvelt Kortwright	
,,	Robert Stordy Dixon	28 Jan. 1842	Charles Fitzgerald Studdert	22 April 1842	William Hamilton (Adjutant)	
,,	..		Philip Antrobus	23 April 1842	Francis John M'Farlane	
,,	..		Alexander Hawtrey	24 April 1842	Lucius John French (Ensign, 13 Dec. 1842)	24 Feb. 1843
,,	..		Wm. Coesvelt Kortwright	25 April 1842	Robert William King	23 June 1843
,,	..		William Hamilton (Adjutant)	29 April 1842	Lord James de Burgh Browne	7 July 1843
,,	..		Francis John M'Farlane	6 Jan. 1843	William Wilberforce Bird	24 Nov. 1843
Paymaster	[W.] Henry Knight		[W.] Henry Knight		Francis Edward Leech (Army, 19 Nov. 1807)	5 Jan. 1844
Adjutant	Robert Cooke		William Hamilton (Cornet, 29 April 1842)	29 April 1842	William Hamilton	
Quarter-master	John Heydon	30 April 1841	Andrew Allan	13 May 1842	Andrew Allan	
Surgeon	Arthur Wood, M.D. (Army, 21 Aug. 1840)	23 July 1841	Arthur Wood, M.D.		Arthur Wood, M.D.	
Assistant-Surgeon	Archibald Stewart (Army, 19 Feb. 1836)	3 Aug. 1841	Archibald Stewart		Archibald Stewart	
,,	..		Walter Geo. Leonard Staunton (Army, 29 May 1840)	22 April 1842	Walter Geo. Leonard Staunton	
Veterinary Surgeon	James Robertson		George Johnston (Army, 26 May 1837)	,,	George Johnston	

Blue. Facings scarlet.

Rank.	1845	Date of Commission.	1846	Date. of Commission.	1847	Date of Commission.
Colonel	[W.] ✠J. W. Sleigh		[W.] ✠J. W. Sleigh		[W.] ✠J. W. Sleigh	
Lieut.-Col.	✠Alexander Campbell, K.H.		✠Alexander Campbell, K.H.		✠A. Campbell, K.H. (Col. 9 Nov. 1846)	
,,	✠John Scott		✠John Scott		✠John Scott (Army, 19 June 1846)	
Major	James Alexander Fullerton, K.H.		J. A. Fullerton, K.H. (Lt.-Col. 25 Apr. 1845)		James Alexander Fullerton, K.H.	
,,	✠Jas. Hope Grant		✠James Hope Grant		✠James Hope Grant	
Captain	Archibald Little		Archibald Little		Archibald Little	
,,	Andrew Spottiswoode		Andrew Spottiswoode		Andrew Spottiswoode	
,,	Edward R. Read		Edward R. Read		Edward R. Read	
,,	Thos. W. M'Mahon		Thos. W. M'Mahon		Thos. W. M'Mahon	
,,	Charles M. Dawson		John Rose H. Rose		John Rose H. Rose	
,,	John Rose H. Rose		Fran. D. Willoughby		James N. Macartney	
,,	Fran. D. Willoughby		James N. Macartney		Kingsmill M. Power	
,,	James Nixon Macartney		Kingsmill Manley Power		Chas. Deacon (Army, 11 July 1845)	20 Mar. 1846
,,	Kingsmill M. Power		Gilbert Thomas Nicholson	24 Oct. 1845	John Cameron Campbell	28 Apr. 1846

NINTH (QUEEN'S ROYAL) LANCERS—*continued.*

Rank.	1845	Date of Commission.	1846	Date of Commission.	1847	Date of Commission.
Lieutenant	Gilbert Thomas Nicholson		John Cameron Campbell		Charles Henry Douglas Donovan	
,,	John C. Campbell		Charles Henry Douglas Donovan		The Hon. Charles Powys	
,,	Charles Henry Douglas Donovan		The Hon. Charles Powys		William Drysdale	
,,	The Hon. C. Powys		William Drysdale		Octavius Henry St. George Anson	
,,	William Drysdale		Octavius Henry St. George Anson		William Wellington Waterloo Humbley	
,,	Octavius Henry St. George Anson		William Wellington Waterloo Humbley		William Richard Newport Campbell	
,,	William Wellington Waterloo Humbley		William Richard Newport Campbell		J. K. Wedderburn	
,,	William Richard Newport Campbell		J. K. Wedderburn		Philip Kemp	
,,	J. K. Wedderburn		Philip Kemp		James Wm. Robarts	
,,	Philip Kemp		James Wm. Robarts		Frederick C. Trower	
,,	Edmund John Turner		Frederick C. Trower		Thomas John Francis	
,,	James W. Robarts		Thomas J. Francis		William F. Anderton (Riding-Master)	
,,	Frederick Courtney Trower		William F. Anderton (Riding-Master)		Philip Antrobus	
,,	Thomas John Francis		Chas. Fitzg. Studdert		Alexander Hawtrey	
,,	William F. Anderton (Riding-Master)		Philip Antrobus		William Coesvelt Kortwright	
,,	Chas. Fitzg. Studdert		Alexander Hawtrey		William Hamilton (Adjutant)	
,,	Philip Antrobus	2 Feb. 1844	William Coesvelt Kortwright	24 Jan. 1845	Francis John M'Farlane	
,,	Alexander Hawtrey	31 May 1844	William Hamilton (Adjutant)		Lucius John French	28 Apr. 1846
,,	..		Francis John M'Farlane	24 Oct. 1845	Henry Nelthorpe (Army, 17 Jan. 1845)	12 June 1846
Cornet	Wm. C. Kortwright		Lucius John French		Robert Wm. King	
,,	William Hamilton (Adjutant)		Robert William King		William Wilberforce Bird	
,,	Francis J. M'Farlane		Lord J. de B. Browne		Chandos F. Clifton	
,,	Lucius John French		William W. Bird		William F. Richards	
,,	Robert William King		Charles Edmund Law		John Henry King	23 Jan. 1846
,,	Lord James de Burgh Browne		Chandos Frederick Clifton		Richard Shaw (Army, 3 Apr. 1846)	8 May 1846
,,	William Wilberforce Bird		William Frederick Richards	24 Jan. 1845	Charles Abbott Delmar	,,
,,	Charles Edmund Law	2 Feb. 1844	..		Richard Hoghton	18 Sept. 1846
,,	Chandos Frederick Clifton	31 May 1844				
Paymaster	Francis E. Leech		Francis E. Leech		Francis E. Leech	
Adjutant	William Hamilton		William Hamilton		William Hamilton	
Quarter-master	Andrew Allan		Andrew Allan		Andrew Allan	
Surgeon	Arthur Wood, M.D.		Arthur Wood, M.D.		Robert J. G. Grant (Army, 2 July 1841)	29 Dec. 1846
Assistant-Surgeon	Archibald Stewart		Archibald Stewart		Archibald Stewart	
	Walter George Leonard Staunton		Walter George Leonard Staunton		Walter George Leonard Staunton	
Veterinary Surgeon	Rd. J. G. Hurford (Army, 17 July 1835)	20 Aug. 1844	Rd. J. Gedaliah Hurford		Rd. J. Gedaliah Hurford	

NINTH (QUEEN'S ROYAL) LANCERS—*continued.*

Rank.	1848	Date of Commission.	1849	Date of Commission.	1850	Date of Commission.
Colonel	[W.] ✠Jas. Wallace Sleigh		[W.] ✠Jas. Wallace Sleigh		[W.] Jas. Wallace Sleigh, C.B.	
Lieut.-Col.	✠ Alex. Campbell, K.H.		✠ Alex. Campbell, K.H.		John Scott, C.B.	
	✠ John Scott		✠ John Scott			
Major	James Alexander Fullerton, K.H.		James Alexander Fullerton, K.H.		James Alexander Fullerton, C.B.	
,,	✠ Jas. Hope Grant		✠ Jas. Hope Grant		Jas. Hope Grant, C.B. (Lt.-Col. 7 June 1849)	
Captain	Archibald Little		Archibald Little		Archibald Little	
,,	Andrew Spottiswoode		Andrew Spottiswoode		Andrew Spottiswoode	
,,	Edward R. Read		Edward R. Read		Edward R. Read	
,,	John R. H. Rose		John R. H. Rose		John R. H. Rose	
,,	John C. Campbell		John C. Campbell		The Hon. Charles Powys	
,,	The Hon. Charles Powys	13 July 1847	The Hon. Charles Powys			
,,	Edward James Pratt (Army, 20 Sept. 1844)	20 July 1847	Edward James Pratt		Edward James Pratt (Major, 7 June 1849)	
,,	Robert A. Yule (Army, 2 July 1847)	,,	Robert A. Yule		Robert A. Yule (Maj. 7 June 1849)	
,,	William Drysdale	29 Oct. 1847	William Drysdale		William Drysdale	
,,	..		..		Octavius Henry St. George Anson	19 Oct. 1849
Lieutenant	Charles Henry Douglas Donovan		Charles Henry Douglas Donovan		Charles Henry Douglas Donovan	
,,	Octavius Henry St. George Anson		Octavius Henry St. George Anson		William Wellington Waterloo Humbley	
,,	William Wellington Waterloo Humbley		William Wellington Waterloo Humbley		William Richard Newport Campbell	
,,	William Richard Newport Campbell		William Richard Newport Campbell		Frederick Courtney Trower	
,,	J. K. Wedderburn		Frederick Courtney Trower		Thomas John Francis	
,,	Philip Kemp		Thomas John Francis		Alexander Hawtrey	
,,	Frederick Courtney Trower		Alexander Hawtrey		William Hamilton (Adjutant)	
,,	Thomas John Francis		William Hamilton (Adjutant)		Francis J. M'Farlane	
,,	Alexander Hawtrey		Francis J. M'Farlane		Lucius John French	
,,	William Hamilton (Adjutant)		Lucius John French		Henry Nelthorpe	
,,	Francis J. M'Farlane		Henry Nelthorpe		Robert Wm. King	
,,	Lucius John French		Robert W. King		John Head	
,,	Henry Nelthorpe		John Head		Wm. Wilberforce Bird	
,,	Robert W. King	17 Mar. 1847	Wm. Wilberforce Bird		John Henry King	
,,	John Head (Army, 19 Nov. 1842)	11 June 1847	Chandos F. Clifton		Charles A. Delmar	
,,	Wm. Wilberforce Bird	6 Aug. 1847	John Henry King		Henry Andrew Sarel	
,,	Chandos F. Clifton	29 Oct. 1847	Charles A. Delmar	23 May 1848	Frederick Ellis	2 Oct. 1849
,,	Wm. Fred. Richards	31 Dec. 1847	George P. Atkinson (Army, 31 March 1848)	1 Aug. 1848	Alfred Eastfield Wilkinson	18 Jan. 1850
,,	John Henry King	,,	Henry A. Sarel	15 Dec. 1848	Henry Scott	22 Feb. 1850
Cornet	Richard Shaw (3 Apr. 1846; Rg.-Master)		Richard Shaw (Riding-Master)		Richard Shaw (Riding-Master)	
,,	Charles A. Delmar		Frederick Ellis		Robert Sutherland	
,,	Richard Hoghton		Edmund D'Arcy Hunt		Alexander James Hardy Elliot	
,,	Henry A. Sarel	13 July 1847	Alfred Eastfield Wilkinson		William Wedderburn Arbuthnot	
,,	Frederick Ellis	13 Aug. 1847	Henry Scott	12 May 1848	Alfred Hives	19 Oct. 1849
,,	Edmund D'Arcy Hunt	29 Oct. 1847	Robert Sutherland	23 May 1848	Archibald Semple Young	23 Nov. 1849
,,	Alfred Eastfield Wilkinson	11 Feb. 1848	Alexander James Hardy Elliot	18 July 1848	John Brinkley	15 Feb. 1850
,,	..		William Wedderburn Arbuthnot	15 Dec. 1848	John George Willis	15 Mar. 1850

NINTH (QUEEN'S ROYAL) LANCERS—*continued.*

Rank.	1848	Date of Commission.	1849	Date of Commission.	1850	Date of Commission.
Paymaster	Francis Edward Leech		Cuthbert Barlow (Army, 3 Jan. 1828; Lieut. 13 Jan. 1814)	21 July 1848	Cuthbert Barlow	
Adjutant	William Hamilton		William Hamilton		William Hamilton	
Qr.-master	Andrew Allan		Andrew Allan		Andrew Allan	
Surgeon	Robert J. G. Grant		Robert J. G. Grant		R. J. G. Grant	
Assistant-Surgeons	W. G. L. Staunton		W. G. L. Staunton		W. G. L. Staunton	
	Eneas Mackintosh Macpherson (Army, 21 Aug. 1840)	12 Feb. 1847	Eneas Mackintosh Macpherson		Eneas Mackintosh Macpherson	
Vet.-Surg.	Rd. J. G. Hurford		Rd. J. G. Hurford		Rd. J. G. Hurford	

Rank.	1851	Date of Commission.	1852	Date of Commission.	1853	Date of Commission.
Colonel	[W.] James Wallace Sleigh, C.B.		[W.] James Wallace Sleigh, C.B.		[W.] James Wallace Sleigh, C.B.	
Lieut.-Col.	John Scott, C.B.		John Scott, C.B.		John Scott, C.B.	
,,	James Hope Grant, C.B.	29 April 1850	James Hope Grant, C.B.		James Hope Grant, C.B.	
Major	Archibald Little	5 April 1850	Archibald Little		Archibald Little	
,,	Andrew Spottiswood	29 April 1850	Wm. Warner Allen (Army, 21 Feb. 1851)	7 Oct. 1851	Wm. Warner Allen	
Captain	John Rose Holden Rose		John Rose Holden Rose		John Rose Holden Rose	
,,	The Hon. Charles Powys		The Hon. Charles Powys		The Hon. Charles Powys	
,,	Edward James Pratt		Edward James Pratt		Edward James Pratt	
,,	Robert Abercromby Yule		Robert Abercromby Yule		Robert Abercromby Yule	
,,	William Drysdale		William Drysdale		William Drysdale	
,,	Octavius Henry St. George Anson		Octavius Henry St. George Anson		Octavius Henry St. George Anson	
,,	Charles H. Douglas Donovan	5 April 1850	William Wellington Waterloo Humbley		William Wellington Waterloo Humbley	
,,	William Wellington Waterloo Humbley	29 April 1850	Augustus Frederick Steele		Augustus Frederick Steele	
,,	Augustus Frederick Steele (Army, 24 May 1850)	21 June 1850	Josias Rogers John Coles (Army, 19 March 1847)	19 Dec. 1851	Josias Rogers John Coles	
,,	..		..		William Richard Newport Campbell	15 Mar. 1853
Lieutenant	William Richard Newport Campbell		William Richard Newport Campbell		Frederick Courtney Trower	
,,	Frederick Courtney Trower		Frederick Courtney Trower		Thomas John Francis	
,,	Thomas John Francis		Thomas John Francis		Alexander Hawtrey	
,,	Alexander Hawtrey		Alexander Hawtrey		William Hamilton (Adjutant)	
,,	William Hamilton (Adjutant)		William Hamilton (Adjutant)		Francis John M'Farlane	
,,	Francis John M'Farlane		Francis John M'Farlane		Lucius John French	
,,	Lucius John French		Lucius John French		Robert William King	
,,	Henry Nelthorpe		Robert William King		John Head	
,,	Robert William King		John Head		John Henry King	
,,	John Head		John Henry King		Henry Andrew Sarel	
,,	William Wilberforce Bird		Henry Andrew Sarel		Frederick Ellis	
,,	John Henry King		Frederick Ellis		Alfred Eastfield Wilkinson	
,,	Chs. A. Delmar		Alfred Eastfield Wilkinson		Henry Scott	
,,	Henry Andrew Sarel		Henry Scott		Richard Shaw (Riding-Master)	

NINTH (QUEEN'S ROYAL) LANCERS—*continued.*

Rank.	1851	Date of Commission.	1852	Date of Commission.	1853	Date of Commission.
Lieutenant	Frederick Ellis		Richard Shaw (Riding-Master)		William Wedderburn Arbuthnot	
,,	Alfred Eastfield Wilkinson		William Wedderburn Arbuthnot	17 June 1851	Archibald Semple Young	
,,	Henry Scott		Alfred Hives	10 Feb. 1852	John George Willis	
,,	Richard Shaw (Riding-Master)	5 April 1850	Archibald Semple Young	27 Feb. 1852	Henry H. Steward	11 June 1852
,,	Hugh Mallett (Army, 28 August 1849	7 Feb. 1851	John George Willis	,,	Robert Charles Kidd	15 Mar. 1853
Cornet	William Wedderburn Arbuthnot		Henry H. Steward		Robert Mills	
,,	Alfred Hives		Robert Charles Kidd		Charles David Rich	
,,	Archibald Semple Young		Robert Mills		Frank Richard Chas. Grant	
,,	John Brinkley		Charles David Rich	17 June 1851	William Cracroft Gordon	
,,	John George Willis		Frank Richard Chas. Grant	16 Sept. 1851	Charles John Heaton Ellis	16 April 1852
,,	Henry H. Steward	12 April 1850	William Cracroft Gordon	13 Feb. 1852	Alfred Stowell Jones	9 July 1852
,,	Robert Charles Kidd	12 July 1850	Christopher Deake Brickmann	12 Mar. 1852	Henry Lavington Payne (Army, 29 July 1852)	27 Aug. 1852
,,	Robert Mills	8 Oct. 1850				
Paymaster	Cuthbert Barlow		Thomas Hanson Ratcliffe (Army, 3 Sept. 1847; Lieut., 16 Aug. 1842)	16 Jan. 1852	Thomas Hanson Ratcliffe	
Adjutant	William Hamilton		William Hamilton		William Hamilton	
Quarter-master	Andrew Allan		Andrew Allan		Peter House	12 Nov. 1852
Surgeon	Robert Joynt Gordon Grant		Robert Joynt Gordon Grant		Robert Joynt Gordon Grant	
Assistant-Surgeon	Eneas Mackintosh Macpherson		William Holmes Jephson, M.D		William Holmes Jephson, M.D.	
,,	William Holmes Jephson, M.D. (Army, 12 July 1844)	15 Nov. 1850	Edward Bailey Tuson (Army, 20 Aug. 1844)	14 Oct. 1851	Edward Bailey Tuson	
Veterinary Surgeon	Richard J. Gedaliah Hurford		Richard J. Gedaliah Hurford		Richard J. Gedaliah Hurford	

Rank.	1854	Date of Commission.	1855	Date of Commission.	1856	Date of Commission.
Colonel	[W.] James Wallace Sleigh, C.B.		[W.] James Wallace Sleigh, C.B. (Gen. 20 June 1854)		[W.] Sir Jas. Wallace Sleigh, K.C.B.	
Lieut.-Col.	John Scott, C.B.		Jas. Hope Grant, C.B. (Col. 28 Nov. 1854)		Jas. Hope Grant, C.B.	
,,	Jas. Hope Grant, C.B.		Archibald Little	20 June 1854	Archibald Little	
,,	..		..		Fred. Chas. Griffiths (Col. 11 Nov. 1851)	25 May 1855
Major	Archibald Little		John Rose Holden Rose	,,	John Rose Holden Rose	
,,	Charles John Foster (Army, 21 Sept. 1852)	Mar. 28 1854	Henry Aime Ouvry (Army, 22 Sept. 1854)	6 Oct. 1854	Henry Aime Ouvry	
Captain	John Rose Holden Rose		The Hon. Charles Powys		The Hon. Charles Powys	
,,	The Hon. Charles Powys		Edward James Pratt		Edward Jas. Pratt	
,,	Edward James Pratt		Robert Abercromby Yule		Robert Abercromby Yule	
,,	Robert Abercromby Yule		William Drysdale		William Drysdale	

NINTH (QUEEN'S ROYAL) LANCERS—*continued.*

Rank.	1854	Date of Commission.	1855	Date of Commission.	1856	Date of Commission.
Captain	William Drysdale		Octavius Henry St. George Anson		Octavius Henry St. George Anson	
,,	Octavius Henry St. George Anson		William Wellington Waterloo Humbley		Augustus Fred. Steele	
,,	William Wellington Waterloo Humbley		Augustus Fred. Steele		Josias Rogers John Coles	
,,	Augustus Fred. Steele		Josias Rogers John Coles		Chardin Philip Johnson	
,,	Josias Rogers John Coles		Chardin Philip Johnson		Anthony M. Fawcett (Army, 12 Jan. 1855)	14 Dec. 1855
,,	Chardin Philip Johnson (Army, 19 Apr. 1850)	6 May 1853	Fred. Courtney Trower	20 June 1854	Thos. Wm. H. Hutchinson (Army, 11 May 1855)	18 Jan. 1856
Lieutenant	Fred. Courtney Trower		Thos. John Francis		William Hamilton (Adjutant)	
,,	Thos. John Francis		Alexander Hawtrey		Lucius John French	
,,	Alexander Hawtrey		William Hamilton (Adjutant)		John Head	
,,	William Hamilton (Adjutant)		Francis John M'Farlane		Henry And. Sarel	
,,	Francis John M'Farlane		Lucius John French		Frederick Ellis	
,,	Lucius John French		John Head		Alfred Eastfield Wilkinson	
,,	Robert Wm. King		John Henry King		Henry Scott	
,,	John Head		Henry And. Sarel		Richard Shaw (Riding-Master)	
,,	John Henry King		Frederick Ellis		John George Willis	
,,	Henry And. Sarel		Alfred Eastfield Wilkinson		Henry Holden Steward	
,,	Frederick Ellis		Henry Scott		Charles David Rich	
,,	Alfred Eastfield Wilkinson		Richard Shaw (Riding-Master)		Robert Mills	
,,	Henry Scott		W. Wedderburn Arbuthnot		Frank Richard Chas. Grant	
,,	Richard Shaw (Riding-Master)		John George Willis		Roger D. Upton (Army, 12 Oct. 1852)	6 April 1855
,,	Wm. Wedderburn Arbuthnot		Henry Holden Steward		Charles John Heaton Ellis	6 July 1855
,,	John George Willis		Charles David Rich		Henry Lavington Payne	14 Sept. 1855
,,	Henry Holden Steward		Robert Mills	20 June 1854	Alf. Stowell Jones	21 Sept. 1855
,,	Chas. Robert Kidd		Frank Richard Chas. Grant	23 June 1854	Robert Blair	2 Nov. 1855
,,	Charles David Rich	23 Dec. 1853	Wm. Cracroft Gordon	1 Dec. 1854	James Goldie	15 Jan. 1856
Cornet	Robert Mills		Chas. John Heaton Ellis		Walter B. Gifford (Army, 2 Oct. 1855)	9 Oct. 1855
,,	Frank Richard Chas. Grant		Alfred Stowell Jones		William Tomkyns	23 Nov. 1855
,,	Wm. Cracroft Gordon		Henry Lavington Payne		Edward King (Ensign, 14 Sept. 1855)	26 Feb 1856
,,	Charles John Heaton Ellis		Robert Blair			
,,	Alfred Stowell Jones		James Goldie	14 July 1854		
,,	Henry Lavington Payne					
,,	Robert Blair (Army, 16 Dec. 1853)	,,				
Paymaster	Thos. Hanson Ratcliffe		Thos. Hanson Ratcliffe		Thos. Hanson Ratcliffe	
Adjutant	William Hamilton		William Hamilton		William Hamilton	
Quarter-master	Peter House		Peter House		Peter House	
Surgeon	Rob. Joynt G. Grant		Rob. Joynt G. Grant		Rob. Joynt G. Grant	
Assistant-Surgeon	Edw. Bailey Tuson		John Jas. Clifford M.D.		John Jas. Clifford, M.D.	
,,	John Chas. Clifford M.D. (Army, 7 Aug. 1846)	27 Jan. 1854	Chris. B. Bassano (Army, 22 Dec. 1848)	9 Feb. 1855	Don. Sinclair Smith (Army, 27 Feb. 1852)	1 Feb. 1856
Veterinary Surgeon	Rd. J. Gedaliah Hurford		Rd. J. Gedaliah Hurford		Rd. J. Gedaliah Hurford	

NINTH (QUEEN'S ROYAL) LANCERS—*continued.*

Rank.	1857	Date of Commission.	1858	Date of Commission.	1859	Date of Commission.
Colonel	[W.] Sir J. Wallace Sleigh, K.C.B.		[W.] Sir J. Wallace Sleigh, K.C.B.		[W.] Sir J. Wallace Sleigh, K.C.B.	
Lieut.-Col.	James Hope Grant, C.B.		Archd. Little, C.B. (Col. 20 June 1857)		Archibald Little, C.B.	
,,	Archibald Little		Henry Aime Ouvry, C.B. (Army, 19 Jan. 1858)	5 Mar. 1858	Henry Aime Ouvry, C.B.	
Major	John Rose Holden Rose		The Hon. C. Powys	18 Sept. 1857	Wm. Drysdale, C.B. (Lt.-Col. 20 July 1858)	
,,	Henry Aime Ouvry		William Drysdale (Army, 19 Jan. 1858)	5 Mar. 1858	Augustus Frederick Steele	22 June. 1858
Captain	The Hon. Charles Powys		Octavius Henry St. George Anson (Maj. 10 Jan. 1858)		Josias Rogers John Coles (Major, 20 July 1858)	
,,	Edward James Pratt (Lieut.-Colonel, 15 Dec. 1856)		Augustus F. Steele		Chardin Philip Johnson (Major, 20 July 1858)	
,,	R. Abercromby Yule (Lieut.-Colonel, 29 Jan. 1857)		Josias Rogers John Coles		Anthony M. Fawcett	
,,	William Drysdale		Chardin P. Johnson		John Head	
,,	Octavius Henry St. George Anson		Anthony M. Fawcett		Frederick Ellis	
,,	Augustus Frederick Steele		Thomas William H. Hutchinson		Alfred Eastfield Wilkinson	
,,	Josias Rogers John Coles		Wm. Hamilton (Maj. 19 Jan. 1858)	9 April 1857	Richard Shaw	22 Mar. 1858
,,	Chardin Philip Johnson		John Head	18 Sept. 1857	Charles David Rich	22 June 1858
,,	Anthony M. Fawcett		Frederick Ellis	11 Oct. 1857	Robert Mills	6 Nov. 1858
,,	Thomas William H. Hutchinson		Alfred Eastfield Wilkinson	5 Mar. 1858		
Lieutenant	William Hamilton		Richard Shaw		Roger D. Upton	
,,	Lucius John French		John George Willis (Riding-Master)		James Goldie	
,,	John Head		Charles David Rich		Arthur Gonne Bell-Martin	
,,	Henry Andrew Sarel		Robert Mills		Alexander Morrogh	
,,	Frederick Ellis		Roger D. Upton		John Evans	
,,	Alfred E. Wilkinson		James Goldie		Piers Thursby	
,,	Henry Scott		Arthur Gonne Bell-Martin		Sam. Ashton Pretor	22 Mar. 1858
,,	Richard Shaw (Riding-Master)		John de Heley Chadwick		Richard F. Thonger (Adjutant)	17 June 1857 (Cornet)
,,	John George Willis		Alexander Morrogh	9 April 1857	T. Stanton Starkey	,,
,,	Charles David Rich		John Evans	21 June 1857	Wm. N. Carleton	6 Nov. 1858
,,	Robert Mills		Piers Thursby	5 Mar. 1858		
,,	Francis R. C. Grant (Adjutant)	23 June 1854				
,,	Roger D. Upton					
,,	Alf. Stowell Jones					
,,	James Goldie					
,,	Arthur Gonne Bell-Martin (Army, 5 June 1855)	18 April 1856				
,,	Edward King	22 July 1856				
,,	John de Heley Chadwick (Army, 14 Sept. 1855)	26 Dec. 1856				
Cornet	Alexander Morrogh	6 June 1856	Sam. Ashton Pretor		Fred. R. Cameron Crofton	
,,	John Evans	22 July 1856	T. Stanton Starkey	15 Jan. 1858	Robert Bury	
,,	Piers Thursby	19 Sept. 1856	Wm. N. Carleton (Army, 2 Feb. 1858)	19 Feb. 1858	Charles Browne Bashford	

NINTH (QUEEN'S ROYAL) LANCERS—*continued.*

Rank.	1857	Date of Commission.	1858	Date of Commission.	1859	Date of Commission.
Cornet	Sam. Ashton Pretor	26 Dec. 1856	Fred. R. Cameron Crofton	26 Feb. 1858	John Russell	13 April 1858
,,	..		Robert Bury	5 Mar. 1858	Charles Agnew	4 June 1858
,,	..		Charles Browne Bashford	30 Mar. 1858	Henry John Hall	2 July 1858
,,	..		..		Oliver Ormerod	17 Sept. 1858
,,	..		..		Henry Angus Bushman	9 Nov. 1858
Paymaster	T. Hanson Ratcliffe		T. Hanson Ratcliffe		Thomas Smales	24 Sept. 1858
Adjutant	Francis R. C. Grant (Lt. 23 June 1854)	20 Nov. 1856	..		Richard F. Thonger (Lt. 13 July 1858)	16 Mar. 1858
Riding-Master	..		..		Charles C. Brooke	15 June 1858
Quarter-master	Peter House		Peter House		Peter House	
Surgeon	H. Franklin (Army, 10 Feb. 1852)	16 Feb. 1856	Henry Franklin		Henry Franklin (Sur. Maj., 1 Oct. 1858)	
Assistant-Surgeon	John James Clifford, M.D.		Sam. Fuller (Army, 23 July 1855)	4 Sept. 1857	Samuel Fuller	
,,	Don. Sinclair Smith		George Evans (Army, 24 Mar. 1854)	26 Jan. 1858	George Evans	
Veterinary Surgeon	Roderick J. Gedaliah Hurford		Frederick Bailey (Army, June 1855)	16 Mar. 1858	Frederick Bailey	

AGENTS—Messrs. Cox and Co.

Rank.	1860	Date of Commission.	1861	Date of Commission.	1862	Date of Commission.
Colonel	[W.] Sir J. Wallace Sleigh, K.C.B.		[W.] Sir J. Wallace Sleigh, K.C.B.		[W.] Sir J. Wallace Sleigh, K.C.B.	
Lieut.-Col.	Archd. Little, C.B.		Archd. Little, C.B.		Wm. Drysdale, C.B.	24 May 1861
Major	Wm. Drysdale, C.B.		Wm. Drysdale, C.B.		Augustus F. Steele	
,,	Augustus F. Steele (Lt.-Col. 26 April 1859)		Augustus F. Steele		Chardin P. Johnson	,,
Captain	Chardin P. Johnson		Chardin P. Johnson		Anthony M. Fawcett	
,,	Anthony M. Fawcett		Anthony M. Fawcett		Frederick Ellis	
,,	Frederick Ellis		Frederick Ellis		Charles David Rich	
,,	Charles David Rich		Charles David Rich (Major, 23 Oct. 1860)		Alexander Morrogh	
,,	R. Stanley Williams Bulkeley (Army, 13 May 1859)	20 Dec. 1859	Alexander Morrogh		Henry Marshall	
,,	James Goldie	21 Feb. 1860	Henry Marshall (Army, 1 July 1859)	17 April 1860	Roger D. Upton	
,,	C. W. Paulet (Army, 12 Nov. 1858)	9 March 1860	Roger D. Upton	2 Feb. 1861	Piers Thursby	,,
,,	Alexander Morrogh	23 Mar. 1860	John Evans	5 Feb. 1861	John Hardy (Army, 3 Dec. 1861)	18 Mar. 1862
Lieutenant	Roger D. Upton		Piers Thursby		Samuel A. Pretor	
,,	Arthur Gonne Bell-Martin		Samuel Ashton Pretor		Richard F. Thonger (Adjutant)	
,,	John Evans		Richard F. Thonger (Adjutant)		T. Stanton Starkey	
,,	Piers Thursby		Thomas Stanton Starkey		W. Macnaughton Erskine	
,,	Samuel Ashton Pretor		Wm. Macnaughton Erskine (Army, 12 Dec. 1859)	25 May 1860	R. Stewart Cleland	
,,	Richard F. Thonger (Adjutant)		R. Stewart Cleland (Army, 8 Apr. 1859)	6 Nov. 1860	E. Barrett Hodges Curteis	

NINTH (QUEEN'S ROYAL) LANCERS—*continued.*

Rank.	1860	Date of Commission.	1861	Date of Commission.	1862	Date of Commission.
Lieutenant	Thomas Stanton Starkey		E. Barrett Hodges Curteis (Army, 23 Nov. 1860)	15 Jan. 1861	Charles Browne Bashford	
,,	Wm. N. Carleton		Charles Browne Bashford	5 Feb. 1861	Wm. G. Blake	
,,	Fred. Rob. Cameron Crofton	21 Feb. 1860	W. G. Blake (Army, 26 Feb. 1858)	15 Feb. 1861	Chas Ashburnham Floyd	1862 24 May
,,	Robert Bury	23 Mar. 1860				
Cornet	Charles Browne Bashford		Charles Agnew		Charles Agnew	
,,	Charles Agnew		Chas. Ashburnham Floyd		J. G. Buchan Hepburn	
,,	Henry John Hall		J. G. Buchan Hepburn		W. Henry Lawrence	
,,	Oliver Ormerod		W. Henry Lawrence	17 April 1860	Bowyer W. Cave-Browne-Cave	
,,	Chas. Ashburnham Floyd	1 July 1859	Bowyer W. Cave-Browne-Cave	29 Jan. 1861	T. Albin Saunders	
,,	J. G. Buchan Hepburn (Ensign, 31 Dec. 1859)	2 March 1860	T. Albin Saunders (Ens., 6 Jan. 1860)	8 Feb. 1861	Perceval Clark	
,,	..		Perceval Clark	22 Feb. 1861	John Chaytor	9 April 1861
,,	..		..		Ed. P. Mackenzie (Army, 18 March 1862)	25 Mar. 1862
Paymaster	Maurice Hartland Mahon (Army, 8 March 1858)	24 May 1859	Maurice Hartland Mahon		Maurice Hartland Mahon	
Adjutant	Richard F. Thonger		Richard F. Thonger		Richard F. Thonger	
Riding-Master	Charles Clements Brooke		Norman Sharp Stevenson	12 Mar. 1861	Norman Sharp Stevenson	
Quarter-master	Peter House		Peter House		Samuel Seggie (Lt. 17 Nov. 1857)	24 May 1861
Surgeon	John. Jas. Clifford, M.D. (Army, 28 Aug. 1857)	5 Aug. 1859	J. J. Clifford, M.D.		J. J. Clifford, M.D.	
Assistant-Surgeon	Samuel Fuller		Geo. C. Clery (Army, 13 Oct. 1854)	12 Oct. 1860	Geo. Carleton Clery	
Veterinary Surgeon	Frederick Bailey		Frederick Bailey		G. Longman (Army, 20 Aug. 1855)	26 Nov. 1861

Rank.	1863	Date of Commission.	1864	Date of Commission.	1865	Date of Commission.
Colonel	[W.] Sir J. Wallace Sleigh, K.C.B.		[W.] Sir J. Wallace Sleigh, K.C.B.		Sir James Hope Grant, G.C.B. (Lieut.-Gen. 3 Oct. 1864)	6 Feb. 1865
Lieut.-Col.	William Drysdale, C.B.		William Drysdale, C.B.		William Drysdale, C.B. (Colonel, 20 June 1864)	
Major	Augustus F. Steele		Augustus F. Steele		Chardin Philip Johnson	
,,	Chardin Philip Johnson		Chardin Philip Johnson		The Hon. Ivo-de-Vesci Twisleton	
,,					Wykeham Fiennes (Army, 20 July 1858)	28 June 1864
Captain	Anthony M. Fawcett		Anthony M. Fawcett		Anthony M. Fawcett	
,,	Frederick Ellis		Charles David Rich		Charles David Rich	
,,	Charles David Rich		Alexander Morrogh		Alexander Morrogh	
,,	Alexander Morrogh		Henry Marshall		Henry Marshall	

NINTH (QUEEN'S ROYAL) LANCERS—*continued.*

Rank.	1863	Date of Commission.	1864	Date of Commission.	1865	Date of Commission.
Captain	Henry Marshall		Piers Thursby		John Hardy	
,,	Piers Thursby		John Hardy		T. Stanton Starkey	
,,	John Hardy		T. Stanton Starkey	5 May 1863	Wm. Macnaughton Erskine	
,,	Samuel Ashton Pretor	24 June 1862	Wm. Macnaughton Erskine	10 July 1863	R. Stewart Cleland	9 Aug. 1864
Lieutenant	Richard Freer Thonger (Adjutant)		Richard Freer Thonger (Adjutant)		Richard Freer Thonger (Adjutant)	
,,	T. Stanton Starkey		R. Stewart Cleland		C. B. Bashford	
,,	Wm. Macnaughton Erskine		C. B. Bashford		Bowyer W. Cave-Browne-Cave	
,,	R. Stewart Cleland		Wm. Greaves Blake		W. Henry Lawrence	
,,	C. B. Bashford		Chas. Ashburnham Floyd		Perceval Clark	
,,	Wm. Greaves Blake		Bowyer W. Cave-Brown-Cave		Charles Agnew	26 Apr. 1864
,,	Chas. Ashburnham Floyd		T. Albin Saunders	5 May 1863	Edward Philippe Mackenzie	26 Apr. 1864
,,	J. G. Buchan Hepburn	24 June 1862	W. Henry Lawrence	29 May 1863	Gilbert C. Stirling	9 Aug. 1864
,,	Bowyer W. Cave-Browne-Cave	16 Jan. 1863	Perceval Clark	10 July 1863	Wm. Savile	21 Mar. 1865
Cornet	Charles Agnew		Charles Agnew		Francis Henry	
,,	Wm. H. Lawrence		John Chaytor		Hon. Oliver George Powlett Montagu	
,,	T. Albin Saunders		Edward Philippe Mackenzie		Hamilton Hamilton Johnston (Army, Ensign, 16 July 1861)	26 Apr. 1864
,,	Perceval Clark		Gilbert Chalmer Stirling		Henry Augustus Candy	26 Apr. 1864
,,	John Chaytor		Wm. Savile		Ronald Mackintosh (Army, Ensign, 29 March 1864)	9 Aug. 1864
,,	Edward Philippe Mackenzie		Francis Henry	30 June 1863	Hon. Edward Fra. John Preston	23 Dec. 1864
,,	Gilbert Chalmer	24 June 1862	The Hon. Oliver G. Powlett Montagu	1 July 1863	Frank de-la-Garde Grissell	21 Mar. 1865
,,	Wm. Savile	16 Jan. 1863	Henry W. Edwards	17 Nov. 1863		
Paymaster	Maurice Hartland Mahon		Maurice Hartland Mahon (Army, Capt. (hon.), 8 Mar. 1863)		Maurice Hartland Mahon	
Adjutant	Richard Freer Thonger		Richard Freer Thonger		Richard Freer Thonger	
Riding-Master	Norman Sharp Stevenson		Norman Sharp Stevenson		Norman Sharp Stevenson	
Quarter-master	Samuel Seggie		Samuel Seggie		Samuel Seggie	
Surgeon	John. Jas. Clifford, M.D.		John Jas. Clifford, M.D.		John Jas. Clifford, M.D.	
Assistant-Surgeon	Geo. Carleton Clery		Geo. Carleton Clery		Geo. Frederick Spry, M.D. (Army, 13 June 1859)	3 June 1864
Vet. Surg.	Geo. Longman		Geo. Longman		Geo. Longman	

NINTH (QUEEN'S ROYAL) LANCERS—*continued.*

Rank.	1866	Date of Commission.	1867	Date of Commission.	1868	Date of Commission.
Colonel	Sir James Hope Grant, G.C.B.		Sir James Hope Grant, G.C.B.		Sir James Hope Grant, G.C.B.	
Lieut.-Col.	Wm. Drysdale, C.B.		Chardin P. Johnson	1 May 1866	Chardin P. Johnson	
Major	Chardin P. Johnson		The Hon. Ivo-de-Vesci Fiennes		The Hon. Ivo-de-Vesci Fienne	
,,	The Hon. Ivo-de-Vesci Fiennes		Charles David Rich (Army, 23 Oct. 1860)	28 Dec. 1866	Charles David Rich	
Captain	Anthony M. Fawcett		Alexander Morrogh		Henry Marshall	
,,	Charles David Rich		Henry Marshall		John Hardy	
,,	Alexander Morrogh		John Hardy		William M. Erskine	
,,	Henry Marshall		William M. Erskine		Robert S. Cleland	
,,	John Hardy		Robert S. Cleland		Bowyer W. Cave-Browne-Cave	
,,	William M. Erskine		Charles B. Bashford		William Savile	20 April 1867
,,	Robert S. Cleland		Bowyer W. Cave-Browne-Cave	1 May 1866	Wm. Clayton Clayton (Army, 9 Mar. 1867)	8 June 1867
,,	Charles B. Bashford	22 Sept. 1865	Perceval Clark	28 Dec. 1866	Henry A. Candy	14 Aug. 1867
Lieutenant	Bowyer W. Cave-Browne-Cave (Adjutant)		William Savile (Adjutant)		Hon. Edward Fra. John Preston	
,,	Perceval Clark		Henry A. Candy		Frank de-la-Garde Grissell (Adjutant)	
,,	William Savile		Hon. Edward Fra. John Preston		Charles Fra. Arnold Howard	
,,	Francis Henry	29 May 1865	Frank de-la-Garde Grissell	1 May 1866	Hugh M'Calmont	
,,	Roger T. Goldsworthy (Army, 16 Aug. 1859)	30 June 1865	Charles Fra. Arnold Howard (Army, 24 Jan. 1865)	17 July 1866	Leslie Aug. Royds Barrow	
,,	Hamilton Hamilton Johnston	18 July 1865	Hugh M'Calmont	27 July 1866	The Hon. Richard Wogan Talbot	
,,	Henry A. Candy	22 Sept. 1865	Leslie Aug. Royds Barrow	14 Sept. 1866	Henry B. Gaskell	
,,	Ronald Mackintosh	8 Dec. 1865	The Hon. Richard Wogan Talbot	28 Dec. 1866	Walter Glyn Lawrell	20 April 1867
,,	The Hon. Edward Fra. John Preston	,,	Henry B. Gaskell	16 Mar. 1867	Charles H. Palairet	14 Aug. 1867
Cornet	Frank de-la-Garde Grissell		John W. L'Estrange Hamilton Browne		Francis A. Herbert	
,,	Hugh M'Calmont	23 June 1865	Walter Glyn Lawrell		Robert J. Barton	
,,	Leslie Aug. Royds Barrow	11 July 1865	Charles H. Palairet		Philip Green	
,,	Hon. Richard Wogan Talbot	12 Sept. 1865	Lord James Hubert Henry Thos. Butler	1 May 1866	Richard St. Leger Moore	20 April 1867
,,	John W. L'Estrange Hamilton Browne	13 Sept. 1865	Francis Alphonse Herbert	27 July 1866	James Alex. Fra. Humberston	
,,	Henry B. Gaskell (Army, 12 Sept. 1865)	20 Oct. 1865	Robert J. Barton	14 Sept. 1866	J. A. F. Stewart-Mackenzie	1 May 1867
,,	Walter Glyn Lawrell	8 Dec. 1865	Philip Green	28 Dec. 1866	Lord Wm. Leslie de la Poer Beresford	8 June 1867
,,	Charles H. Palairet	12 Dec. 1865	..		The Hon. G. E. P. Willoughby	12 June 1867
,,	..		..		William Percy	14 Aug. 1867
Paymaster	Maurice H. Mahon		Maurice H. Mahon		Maurice H. Mahon	
Adjutant	B. W. Cave-Browne-Cave	30 June 1865	William Savile	1 May 1866	F. de la G. Grissell	20 April 1867
Riding-Master	Norman S. Stevenson		John Crowdy	2 Nov. 1866	John Crowdy	
Quarter-master	Samuel Seggie		Samuel Seggie		Samuel Seggie	
Surgeon	John James Clifford, M.D.		John James Clifford, M.D. (Surg.-Maj. 7 Aug. 1866)		John James Clifford, M.D.	
Assistant-Surgeon	Geo. Frederick Spry, M.D.		Maximilian Grant, M.D. (Army, 10 Mar. 1857)	18 Jan. 1867	Maximilian Grant, M.D.	
Veterinary Surgeon	George Longman		George Longman		G. Longman (Vet. Surg. 1st Class, 21 Aug. 1867)	

NINTH (QUEEN'S ROYAL) LANCERS—*continued.*

Rank	1869	Date of Commission.	1870	Date of Commission.	1871	Date of Commission.
Colonel	Sir Jas. Hope Grant, G.C.B.		Sir Jas. Hope Grant, G.C.B.		Sir Jas. Hope Grant, G.C.B.	
Lieut.-Col.	The Hon. Ivo-de-Vesci Fiennes	19 May 1869	The Hon. Ivo-de-Vesci Fiennes		The Hon. Ivo-de-Vesci Fiennes	
Major	Henry Marshall	11 April 1868	Henry Marshall		Henry Marshall	
,,	John Hardy	19 May 1869	John Hardy		John Hardy	
Captain	Robert Stewart Cleland		Robert Stewart Cleland		Robert Stewart Cleland	
,,	Bowyer W. Cave-Browne-Cave		Bowyer W. Cave-Browne-Cave		Bowyer W. Cave-Browne-Cave	
,,	Wm. Savile		Wm. Clayton Clayton		Wm. Clayton Clayton	
,,	Wm. Clayton Clayton		The Hon. E. F. John Preston		Frank de-la-Garde Grissell	
,,	Hen. Aug. Candy		Frank de-la-Garde Grissell		Hen. B. Gaskell	
,,	The Hon. E. F. John Preston	1 April 1868	Hen. B. Gaskell	6 July 1870	Walter Glyn Lawrell	
,,	Frank de-la-Garde Grissell	19 May 1869	Sir Morison Barlow, Bt.		Chas. Harvey Palairet	25 Mar. 1871
,,	Hugh M'Calmont	29 May 1869	Walter Glyn Lawrell	27 July 1870	Fra. Alphonse Herbert	25 Mar. 1871
Lieutenant	C. F. A., Earl of Wicklow, *s.*	24 Jan. 1865	The Hon. R. Wogan Talbot		The Hon. R. Wogan Talbot	
,,	The Hon. R. Wogan Talbot		Chas. Harvey Palairet		Robert Johnston Barton, *s.*	
,,	Hen. Brooks Gaskell		Fra. Alphonse Herbert		Philip Green	
,,	Walter Glyn Lawrell		Robert Johnston Barton, *s.*		Richard St. Leger Moore	
,,	Chas. Harvey Palairet		Philip Green		J. A. F. H. Stewart-Mackenzie	
,,	Fra Alphonse Herbert	11 April 1868	Richard St. Leger Moore		Lord W. L. de la P. Beresford	
,,	Robert Johnston Barton,	2 May 1868	J. A. F. H. Stewart-Mackenzie	8 June 1870	The Hon. G. E. P. Willoughby	
,,	Philip Green	19 May 1869	Lord W. L. de la P. Beresford	6 July 1870	William Percy, Adjt.	25 Mar. 1871
,,	Richard St. Leger Moore	29 May 1869	The Hon. G. E. P. Willoughby	27 July 1870	Alf. L. Wheeler	,,
,,	..		..		Wm. Hen. Fife	,,
Cornet	J. A. F. H. Stewart-Mackenzie		William Percy		Strange Gould Butson	
,,	Lord W. L. de la P. Beresford		Alf. L. Wheeler			
,,	The Hon. G. E. P. Willoughby		Wm. Hen. Fife			
,,	William Percy		Strange Gould Butson	27 July 1870		
,,	Alf. Llewelyn Wheeler	11 April 1868				
,,	Wm. Henry Fife	1 April 1868				
Paymaster	Maurice H. Mahon		Maurice H. Mahon		Maurice H. Mahon	
Adjutant	W. Percy	19 May 1869	W. Percy		W. Percy	
Riding-Master	John Crowdy		John Crowdy		John Crowdy	
Quarter-master	Geo. Fuller	30 Oct. 1869	Geo. Fuller		Geo. Fuller	
Surgeon	John J. Clifford, M.D.		John J. Clifford, M.D.		John J. Clifford, M.D.	
Assistant-Surgeon	Max. Grant, M.D.		Max. Grant, M.D.		J. V. T. Malcolm, M.D.	31 Mar 1864
Veterinary Surgeon	Wm. C. Lord. (1st Class, 1 July 1859)	26 Jan. 1869	E. T. Cheesman	20 Nov. 1867	E. T. Cheesman	

Blue. Facings Scarlet. Plume—Black and White.

AGENTS—Messrs Cox & Co. IRISH AGENTS—Sir E. R. Borough, Bart., Armit & Co.

NINTH (QUEEN'S ROYAL) LANCERS—*continued.*

Rank.	1872	Date of Commission.	1873	Date of Commission.	1874	Date of Commission.
Colonel	Sir James Hope Grant, G.C.B.		Sir James Hope Grant, G.C.B.		Sir James Hope Grant, G.C.B.	
Lieut.-Col.	The Hon. W. Ivo-de-Vesci Fiennes		The Hon. Ivo de Vesci T. W. Fiennes		Henry Marshall	27 June 1874
Major	Henry Marshall		Henry Marshall		John Hardy	
,,	John Hardy		John Hardy			
Captain	Robert Stewart Cleland		Robert Stewart Cleland		Robert Stewart Cleland	
,,	Bowyer W. Cave-Browne-Cave		Bowyer W. Cave-Browne-Cave		Wm. Clayton-Clayton	
,,	Wm. Clayton-Clayton		Wm. Clayton-Clayton		Frank de-la-Garde Grissell	
,,	Frank de-la-Garde Grissell		Frank de-la-Garde Grissell		Hen. Aug. Bushman	
,,	Hen. B. Gaskell		Hen. B. Gaskell		Henry R. Abadie	
,,	Charles Harvey Palairet		Hen. Aug. Bushman		The Hon. Heneage Legge (Army, 11 Dec. 1866)	10 Dec. 1873
,,	Edward Williams	10 June 1871	Robert Johnston Barton, *s.*	26 Mar. 1873	Philip Green	25 Feb. 1874
,,	Hen. Aug. Bushman	9 Mar. 1872	Henry R. Abadie (Army, 14 Feb. 1872)	16 Aug. 1873	Geo. Wm. Hayne (Army, 9 Sept. 1864)	9 Sept. 1874
Lieutenant	Robert Johnston Barton, *s.*		Philip Green		Richard St. Leger Moore	
,,	Philip Green		Richard St. Leger Moore		J. A. F. H. Stewart-Mackenzie	
,,	Richard St. Leger Moore		J. A. F. H. Stewart-Mackenzie		Lord W. L. de la P. Beresford	
,,	J. A. F. H. Stewart-Mackenzie		Lord W. L. De la P. Beresford		The Hon. G. E. P. Willoughby	
,,	Lord W. L. de la P. Beresford		The Hon. G. E. P. Willoughby		William Percy (Adjutant)	
,,	Hon. G. E. P. Willoughby		William Percy (Adjutant)		Alfred L. Wheeler	
,,	William Percy (Adjutant)		Alfred L. Wheeler		Wm. Hen. Fife	
,,	Alfred L. Wheeler		Wm. Hen. Fife		Strange Gould Butson	
,,	Wm. Hen. Fife		Strange Gould Butson		James J. Scott-Chisholme	
,,	Strange Gould Butson	28 Oct. 1871	Bloomfield Gough (Army, 28 Oct. 1871)	23 April 1873	Bloomfield Gough	
,,	..		..		The Hon. Geo. W. G. Russell	28 Oct. 1873
,,	..		..		Egremont E. Shearburn	23 May 1874
Sub.-Lieut.	The Hon. Geo. W. G. Russell	28 Oct. 1871	The Hon. Geo. W. G. Russell		The Hon. Eustace Vesey	
,,	Edward G. B. Tribe	31 Oct. 1871	James J. Scott-Chisholme		Geo. A. Penrhys Evans	13 June 1874
,,	James J. Scott-Chisholme	10 Jan. 1872	The Hon. D. W. W. Westenra	4 Dec. 1872		
,,	..		The Hon. Eustace Vesey	23 April 1873		
Paymaster	Maurice H. Mahon		Maurice H. Mahon, *m.*		Maurice H. Mahon, *m.*	
Adjutant	W. Percy		W. Percy		W. Percy	
Riding-Master	John Crowdy		John Crowdy		John Crowdy	
Quarter-master	George Fuller		George Fuller		George Fuller	
Surgeon	R. Lewer	9 Sept. 1871	R. Lewer (Surgeon-Major)	2 Dec. 1871	R. Lewer (Surgeon-Major)	
Assistant-Surgeon	J. V. T. Malcolm, M.D.					
Veterinary Surgeon	E. T. Cheesman		E. T. Cheesman		E. T. Cheesman	

NINTH (QUEEN'S ROYAL) LANCERS—*continued.*

Rank.	1875	Date of Commission.	1876	Date of Commission.	1877	Date of Commission.
Colonel	Sir Archibald Little, K.C.B. (Maj.-Gen. 6 Mar. 1868)	8 Mar. 1875	Sir Archibald Little, K.C.B.		Sir Archibald Little, K.C.B. (Lt.-Gen.)	
Lieut.-Col.	Henry Marshall		Henry Marshall		Henry Marshall	
Major	John Hardy		John Hardy		John Hardy	
Captains	Robert S. Cleland, *d.*		Robert S. Cleland, *d.*		R. S. Cleland, *m.*	
,,	W. Clayton-Clayton		W. Clayton-Clayton		H. Aug. Bushman, *d.*	
,,	Hen. Aug. Bushman		Hen. Aug. Bushman		Henry R. Abadie	
,,	Henry R. Abadie		Henry R. Abadie		The Hon. Heneage Legge	
,,	The Hon. Heneage Legge		The Hon. Heneage Legge		Herbert W. Apperley	
,,	Richard Chambers	23 Jan. 1875	Herbert W. Apperley		J. A. F. H. Stewart-Mackenzie	
,,	Herbert W. Apperley (Army, 9 Sept. 1874)	1 Dec. 1875	J. A. F. H. Stewart-Mackenzie	15 July 1876	Lord W. L. de la P. Beresford, *s.*	27 Dec. 1876
Lieutenant	J. A. F. H. Stewart-Mackenzie		Lord W. L. de la P. Beresford, *s.*		The Hon. G. E. P. Willoughby, *d.*	
,,	Lord W. L. de la P. Beresford		The Hon. G. E. P. Willoughby, *d.*		William Hen. Fife	
,,	The Hon. G. E. P. Willoughby, *d.*		William Hen. Fife		Strange Gould Butson	
,,	W. Percy (Adjutant)		Strange Gould Butson		J. J. Scott-Chisholme	
,,	William Hen. Fife		J. J. Scott Chisholme		Bloomfield Gough	
,,	Strange Gould Butson		Bloomfield Gough		The Hon. Eustace Vesey (Adjutant)	
,,	J. J. Scott-Chisholme		The Hon. Eustace Vesey (Adjutant)	24 July 1875	Egremont E. Sherburn	
,,	Bloomfield Gough		Egremont E. Shearburn		George A. Penrhys Evans	
,,	The Hon. Eustace Vesey		George A. Penrhys Evans	13 June 1874	The Hon. Chas. Geo. Lascelles	
,,	Egremont E. Shearburn		The Hon. Chas. G. Lascelles		Chas. J. W. Trower	
,,	The Hon. Charles George Lascelles (Army, 27 Mar. 1872)	25 Nov. 1874	Chas. J. W. Trower		Charles John R. Hearsay	11 Feb. 1875
,,	Chas. J. W. Trower	2 Dec. 1874	Ed. B. M'Innis		Ed. B. M'Innis	
,,	Edward B. M'Innis Army, 29 Apr. 1874)	23 Oct. 1875	Henry Castell Hopkins	26 Aug. 1876	H. Castell Hopkins	
Sub.-Lieut.	G. A. Penrhys Evans		Chas. J. R. Hearsey		C. J. B. H. Dressner	
,,	Chas. J. R. Hearsey	11 Feb. 1875	C. J. B. H. Dressner	11 Sept. 1876	..	
Paymaster	John W. M'Farlan (Army, 16 Oct. 1866; Capt. (hon.), 16 Oct. 1871)	24 Dec. 1874	John W. M'Farlan (Hon. Capt.)		John W. M'Farlan (Hon. Capt.)	
Adjutant	W. Percy		The Hon. E. Vesey	24 July 1875	The Hon. E. Vesey	
Riding-Master	John Marshall	27 Feb. 1875	John Marshall		John Marshall	
Qr.-master	George Fuller		George Fuller		George Fuller	
Medical Officer	R. Lewer (Sur.-Maj.)	2 Dec. 1871	R. Lewer (Sur.-Maj.)		R. Lewer (Sur.-Maj.)	
Veterinary Surgeon	J. P. Adams	16 Dec. 1874	J. P. Adams		Jas. Aug. Woods	5 Sept. 1877

NINTH (QUEEN'S ROYAL) LANCERS—*continued.*

Rank.	1878	Date of Commission.	1879	Date of Commission.	1880	Date of Commission.
Colonel	Sir A. Little, K.C.B.		Sir A. Little, K.C.B.		Sir A. Little, K.C.B. (Gen. 21 Jan. 1880)	
Lieut.-Col.	Henry Marshall		Rob. S. Cleland	27 June 1879	Henry Aug. Bushman	8 Aug. 1880
Major	Rob. S. Cleland		Hen. Aug. Bushman	,,	Henry R. Abadie	22 Dec. 1879
Captain	Henry A. Bushman, *d.*		Hen. R. Abadie, *m.*		The Hon. H. Legge, *m.*	
,,	Henry R. Abadie		The Hon. H. Legge		Herbert W. Apperley	
,,	The Hon. H. Legge		Herbert W. Apperley		J. A. F. H. Stewart-Mackenzie	
,,	Herbert W. Apperley		J. A. F. H. Stewart-Mackenzie		Lord W. L. de la P. Beresford, V.C.	
,,	J. A. F. H. Stewart-Mackenzie		Lord W. L. de la P. Beresford, V.C., *s.*		Wm. Hen. Fife	
,,	Lord W. L. de la P. Beresford, *s.*		Wm. Hen. Fife		Jas. J. Scott-Chisholme	
,,	Wm. Henry Fife	16 Feb. 1878	Strange G. Butson		Bloomfield Gough	
,,	Strange G. Butson	,,	J. J. Scott-Chisholme		The Hon. E. Vesey	
,,	J. J. Scott-Chisholme	2 March 1878	Bloomfield Gough		Geo. A. Penrhys Evans	8 Aug. 1880
,,	Bloomfield Gough	17 July 1878	The Hon. Eustace Vesey	27 June 1879	Chas. J. W. Trower	10 Nov. 1880
Lieutenant	The Hon. Eustace Vesey, *d.*		Egre. E. Shearburn		Edward B. M'Innis (Adjutant)	
,,	Egre. E. Shearburn		Geo. A. P. Evans		Henry C. Hopkins	
,,	Geo. A. P. Evans		The Hon. Chas. Geo. Lascelles		Chas. J. B. H. Dressner (prob.)	
,,	The Hon. Chas. Geo. Lascelles		Chas. J. W. Trower		John Hen. Lamont	
,,	Chas. J. W. Trower		Chas. J. R. Hearsey		James Hunter	12 Dec. 1879
,,	Chas. J. R. Hearsey		Edward B. M'Innis (Adjutant)		Edw. H. de T. G. Bell-Martin	1 Jan. 1880
,,	Edward B. M'Innis (Adjutant)		Henry C. Hopkins		Malcolm Orme Little	25 Feb. 1880
,,	Henry C. Hopkins		Chas. J. B. H. Dressner (prob.)		Edward P. Johnson	8 Aug. 1880
,,	Chas. J. B. H. Dressner	11 Sept. 1876	John Hen. Lamont	29 Nov. 1877	Charles Bishop	10 Nov. 1880
Sub-Lieut.	John Hen. Lamont	29 Nov. 1876				
2nd Lieut.	Wm. Percy Ricardo	13 Mar. 1878	Wm. Percy Ricardo		Harry Ritchie Alston (Army, 13 Oct. 1877)	7 Jan. 1880
,,	James Hunter	20 Mar. 1878	James Hunter		Cyril St. C. Cameron (Army, 22 Mar. 1879)	,,
,,	Edw. H. de T. G. Bell-Martin	,,	Edw. H. de T. G. Bell-Martin		Hon. Richard H. Westenra	24 Jan. 1880
,,	Malcolm Orme Little (Army, 11 May 1878)	19 Oct. 1878	Malcolm Orme Little		Ronald M'Farlane	31 Jan. 1880
,,	..		Edward P. Johnson	23 July 1879	Walter K. W. Jenner	16 Oct. 1800
,,	..		Charles Bishop	13 Aug. 1879	Forrester F. Colvin	23 Oct. 1880
Paymaster	John W. M'Farlan (Hon. Capt.)		John W. M'Farlan (Hon. Capt.)		John W. Macfarlan (Hon. Capt.)	
Adjutant	E. B. M'Innis	22 June 1877	E. B. M'Innis		E. B. M'Innis	
Riding-Master	John Marshall		John Marshall		John Marshall	
Quarter-master	Geo. Fuller		John Forsythe	27 Sept. 1879	John Forsythe	
Vet. Surg.	Jas. Augs. Woods		Jas. Augs. Woods		Jas. Augs. Woods	

NINTH (QUEEN'S ROYAL) LANCERS—*continued.*

Rank.	1881	Date of Commission.	1882	Date of Commission.	1883	Date of Commission.
Colonel	Sir A. Little, K.C.B.		Sir A Little, K.C.B.		Sir A. Little, K.C.B.	
Lieut.-Col.	Henry A. Bushman		H. A. Bushman, C.B.		H. A. Bushman, CB.	
"	..		..		H. R. Abadie	
Major	H. R. Abadie		H. R. Abadie (Lt.-Col. 1 July 1881)		The Hon. H. Legge	
"	..		The Hon. H. Legge (Brev. 29 Apr. 1880)	1 July 1881	H. W. Apperley	
"	..		H. W. Apperley (Brev. 2 Mar. 1881)	"	J. A. Stewart-Mackenzie	
"	..		J. A. Stewart Mackenzie (Brev. 2 Mar. 1881)	"		
Captain	The Hon. H. Legge		Lord Wm. Beresford, V.C.		Lord Wm. Beresford, V.C.	
"	H. W. Apperley		W. H. Fife		W. H. Fife	
"	J. A. F. H. Stewart-Mackenzie		J. J. Scott-Chisholme (Brev.-Maj. 2 Mar. 1881)		J. Scott-Chisholme (Adj. 18 Nov. 1882)	
"	Lord William de la P. Beresford, V.C.		B. Gough (Brev.-Maj. 2 Mar. 1881)		B. Gough	
"	W. H. Fife		The Hon. E. Vesey		The Hon. E. Vesey	
"	J. J. Scott-Chisholme		G. A. P. Evans		G. A. P. Evans	
"	B. Gough		C. J. Trower		C. J. Trower	
"	The Hon. E. Vesey		E. B. M'Innis	1 July 1881	E. B. M'Innis	
"	G. A. P. Evans					
"	Chas. J. W. Trower					
Lieutenant	E. B. M'Innis (Adj.)		H. C. Hopkins		H. C. Hopkins	
"	H. C. Hopkins		J. H. Lamont		The Hon. C. G. Lascelles (reappointed from half-pay) (Army, 27 Mar. 1872)	
"	C. J. B. H. Dressner		J. Hunter		J. H. Lamont	
"	J. H. Lamont		E. H. Bell-Martin		M. O. Little	
"	James Hunter		M. O. Little		E. P. Johnson	
"	Ed. H. Bell-Martin		E. P. Johnson		C. Bishop	
"	M. O. Little		C. Bishop		C. St. C. Cameron	
"	E. P. Johnson		H. R. Alston		R. M'Farlane	
"	C. Bishop		C. St. C. Cameron		W. K. Jenner	
"	..		R. M'Farlane		F. F. Colvin	
"	..		W. K. Jenner		F. W. Duff	
"	..		F. F. Colvin		F. H. Reynard (Army, 10 Sept. 1875)	18 Nov. 1882
"	..		F. W. Duff (2nd Lt. 19 Feb. 1881)	"		
2nd Lieut.	H. R. Alston					
"	C. St. C. Cameron					
"	Hon. R. H. Westenra					
"	R. M'Farlane					
"	W. K. W. Jenner					
"	F. F. Colvin					
Paymaster	John W. M'Farlan (Hon. Capt.)		J. M'Farlan (Hon. Capt.)		J. M'Farlan (Hon. Capt.)	
Riding-Master	John Marshall		J. Marshall		J. Marshall	
Qr.-master	John Forsythe		J. Forsythe		J. Forsythe	
Vet. Surg.	J. A. Woods					

NINTH (QUEEN'S ROYAL) LANCERS—*continued.*

Rank.	1884	Date of Commission.	1885	Date of Commission.	1886	Date of Commission.
Colonel	Sir A. Little, K.C.B.		Sir A. Little, K.C.B.		Sir A. Little, K.C.B.	
Lt.-Col.	H. A. Bushman, C.B.		H. A. Bushman, C.B.		H. R. Abadie (Col. 1 July 1885)	
,,	H. R. Abadie (Com. Regt. 15 Dec. 1884)		H. R. Abadie		Hon. H. Legge	
Major	The Hon. H. Legge (Lt.-Col. 15 Dec. 1884)		The Hon. H. Legge		H. W. Apperley	
,,	H. W. Apperley		H. W. Apperley		J. A. Stewart Mackenzie	
,,	J. A. Stewart-Mackenzie		J. A. Stewart-Mackenzie		Lord W. Beresford, V.C.	15 Dec. 1884
,,	..		..		J. J. Scott-Chisholme	,,
Captain	Lord Wm. Beresford, V.C.		Lord W. Beresford, V.C.		B. Gough	
,,	J. Scott-Chisholme		J. Scott-Chisholme		The Hon. E. Vesey	
,,	B. Gough		B. Gough		G. A. P. Evans	
,,	The Hon. E. Vesey		The Hon. E. Vesey		E. B. M'Innis	
,,	G. A. P. Evans		G. A. P. Evans		J. H. Lamont	23 Nov. 1885
,,	E. B. M'Innis		E. B. M'Innis			
Lieutenant	H. C. Hopkins		H. C. Hopkins		M. O. Little	
,,	J. H. Lamont		J. H. Lamont		E. P. Johnson	
,,	M. O. Little		M. O. Little		C. Bishop	
,,	E. P. Johnson		E. P. Johnson		C. St. C. Cameron	
,,	C. Bishop		C. Bishop		R. M'Farlane	
,,	C. St. C. Cameron		C. St. C. Cameron		W. K. Jenner	
,,	R. M'Farlane		R. M'Farlane		F. F. Colvin	
,,	W. K. Jenner		W. K. Jenner		F. W. Duff	
,,	F. F. Colvin		F. F. Colvin		F. H. Reynard	
,,	F. W. Duff		F. W. Duff		The Hon. Claude Willoughby	12 Nov. 1884
,,	F. H. Reynard		F. H. Reynard (Adj. 20 May 1885)		The Hon. L. King-Noel	29 Aug. 1885
,,	Ralph Broughton	10 Mar. 1883	R. Broughton		Lord D. Compton	2 Sept. 1885
Paymaster	..		J. C. S. Irving		E. H. Maunsell (Hon. Maj. 26 Aug. 1884)	
Riding-Master	J. Marshall		J. Marshall		J. Marshall (Hon. Capt. 27 Feb. 1885)	
Qr.-master	J. Forsythe		J. Forsythe		J. Forsythe	

Rank.	1887	Date of Commission.	1888	Date of Commission.	1889	Date of Commission.
Colonel	Sir A. Little, K.C.B.		Sir A. Little, K.C.B.		Sir A. Little, K.C.B.	
Lieut.-Col.	H. R. Abadie		The Hon. H. Legge (Com. Regt.)	1 July 1887	The Hon. H. Legge	
,,	The Hon. H. Legge					
Major	W. H. Apperley		W. H. Apperley		J. A. Stewart-Mackenzie	
,,	J. A. Stewart-Mackenzie		J. A. Stewart-Mackenzie			
,,	Lord Wm. Beresford, V.C.		Lord Wm. Beresford, V.C.		Lord Wm. Beresford, V.C.	
,,	J. Scott-Chisholme		J. Scott-Chisholme		J. Scott-Chisholme	
,,	..		B. Gough	8 Dec. 1886	B. Gough	
Captain	B. Gough		J. H. Lamont		J. H. Lamont	
,,	E. B. M'Innis		F. H. Reynard		F. H. Reynard	
,,	J. H. Lamont		M. O. Little		M. O. Little	
,,	F. H. Reynard (Adjt. 20 Oct. 1886)	7 Aug. 1886	E. P. Johnson		C. Bishop C. St. C. Cameron	
,,	M. O. Little	Aug. 1886	C. Bishop	8 Dec. 1886	R. M'Farlane	23 May 1888
,,	E. P. Johnson		C. St. C. Cameron	30 Mar. 1887	W. K. Jenner	,,
,,	..		..		F. F. Colvin	10 Aug. 1888

NINTH (QUEEN'S ROYAL) LANCERS—*continued.*

Rank.	1887	Date of Commission.	1888	Date of Commission.	1889	Date of Commission.
Lieutenant	C. Bishop		R. M'Farlane		F. W. Duff	
,,	C. St. C. Cameron		W. K. Jenner (Adjt. 30 July 1887)		The Hon. C. Willoughby	
,,	R. M'Farlane		F. F. Colvin		The Hon. L. King-Noel	
,,	W. K. Jenner		F. W. Duff		Lord Douglas Compton	
,,	F. F. Colvin		The Hon. C. Willoughby		The Hon. Victor Pelham	
,,	F. W. Duff		The Hon. L. King-Noel		H. F. Stanley	21 Nov. 1888
,,	The Hon. C. Willoughby		Lord Douglas Compton		H. E. W. Fiennes	,,
,,	The Hon. L. King-Noel		The Hon. Victor Pelham		J. L. de Laski	,,
,,	Lord Douglas Compton					
,,	The Hon. Victor Anderson Pelham	6 Nov. 1886				
,,	Lord John Thynne	29 Dec. 1886				
2nd Lieut.	..		Herbert F. W. Stanley	26 Feb. 1887	G. F. Henry	
,,	..		Henry E. J. Wykeham Fiennes	30 April 1887	George Paget Ellison	22 Aug. 1888
,,	..		J. L. A. de Laski	14 Sept. 1887		
,,	..		Gilbert Frank Henry	16 Nov. 1887		
Paymaster	E. H. Maunsell		E. H. Maunsell		A. M. Leslie (Captain, Suffolk Regiment)	
Riding-Master	J. Marshall		E. H. Nicholson (Hon. Lieut. 24 May 1879)		E. H. Nicholson	
Quarter-master	J. Forsythe		J. Forsythe		J. Forsythe	

Rank.	1890	Date of Commission.	1891	Date of Commission.	1892	Date of Commission.
Colonel	Sir A. Little, G.C.B.		Sir A. Little, G.C.B.		W. Drysdale, C.B. (Lieut.-General, 11 June 1891)	
Lieut.-Col.	The Hon. H. Legge (Colonel)	15 Dec. 1888	J. A. Stewart-Mackenzie (Com. Regt.)	15 Dec. 1890	J. A. Stewart-Mackenzie	
Major	J. A. Stewart-Mackenzie		B. Gough		B. Gough	
,,	Lord W. Beresford, V.C.		H. C. Morland		H. C. Morland	
,,	B. Gough		..		J. H. Lamont	15 Dec. 1890
,,	H. C. Morland (Army, Nov. 1888)					
Captain	J. H. Lamont		J. H. Lamont		F. H. Reynard	
,,	F. H. Reynard		F. H. Reynard		M. O. Little	
,,	M. O. Little		M. O. Little		C. Bishop	
,,	C. Bishop		C. Bishop		C. St. C. Cameron	
,,	C. St. C. Cameron		C. St. C. Cameron		R. M'Farlane	
,,	R. M'Farlane		R. M'Farlane		W. K. Jenner	
,,	W. K. Jenner		W. K. Jenner		F. F. Colvin	
,,	F. F. Colvin		F. F. Colvin		F. W. Duff	
,,	F. W. Duff	10 March 1889	F. W. Duff		The Hon. C. Willoughby	
,,	The Hon. C. Willoughby	4 Dec. 1889	The Hon. C. Willoughby			
Lieutenant	The Hon. L. King-Noel		The Hon. L. King-Noel		The Hon. L. King-Noel (Adjutant, 31 Aug. 1892)	

NINTH (QUEEN'S ROYAL) LANCERS—*continued.*

Rank.	1890	Date of Commission.	1891	Date of Commission.	1892	Date of Commission.
Lieutenant	Lord D. Compton		Lord D. Compton		Lord D. Compton	
,,	The Hon. Victor Pelham		The Hon. V. Pelham (Adj., 30 July 1891)		The Hon. Victor Pelham	
,,	H. F. Stanley		H. F. Stanley		H. F. Stanley	
,,	H. E. W. Fiennes		H. E. W. Fiennes		H. E. W. Fiennes	
,,	J. L. de Laski		G. F. Henry		G. F. Henry	
,,	..		G. P. Ellison	4 Jan. 1890	G. P. Ellison	
,,	..		..		D. G. Campbell	1890 9 Dec.
,,	..		..		Lord C. Bentinck	1891 23 Jan.
,,	..		..		W. von Schröder	1 Oct.
2nd Lieut.	G. P. Ellison		D. G. Campbell		James W. Harold Russel	4 March
,,	David Graham M. Campbell	1889 16 March	Lord C. Bentinck		Desmond J. E. Beale-Browne	30 May
,,	Lord Chas. Bentinck	8 June	W. von Schröder		Ernest Fitzroy Bell	5 Dec.
,,	W. H. von Schröder	4 Dec.				
Paymaster	A. M. Leslie					
Riding-Master	E. H. Nicholson		E. H. Nicholson		E. H. Nicholson	
Qr.-master	J. Forsythe		J. Forsythe		J. Forsythe	

Rank.	1893	Date of Commission.	1894	Date of Commission.	1895	Date of Commission.
Colonel	Lieut.-General Wm. Drysdale, C.B.		Lieut.-General Wm. Drysdale, C.B.		Sir Wm. Drysdale, K.C.B.	
Lieut.-Col.	J. A. Stewart-Mackenzie		J. A. Stewart-Mackenzie		J. A. Stewart-Mackenzie (Colonel)	15 Dec. 1894
Major	B. Gough		B. Gough		B. Gough	
,,	H. C. Morland		H. C. Morland		H. C. Morland	
,,	J. H. Lamont		J. H. Lamont		M. O. Little	5 Sept. 1894
Captain	M. O. Little		M. O. Little		C. Bishop	
,,	C. Bishop		C. Bishop		W. K. Jenner	
,,	C. St. C. Cameron		C. St. C. Cameron		F. F. Colvin	
,,	W. K. Jenner		W. K. Jenner		F. W. Duff	
,,	F. F. Colvin		F. F. Colvin		The Hon. C. Willoughby	
,,	F. W. Duff		F. W. Duff		F. T. Lund	
,,	The Hon. C. Willoughby		The Hon. C. Willoughby		The Hon. L. King-Noel	
,,	Frederick T. Lund	11 May 1892	F. T. Lund		Lord Douglas Compton	,,
,,	The Hon. L. King-Noel (Adjutant)	31 Aug. 1892	The Hon. L. King-Noel		H. F. Stanley	,,
,,	..		..		H. E. W. Fiennes	17 Sept. 1894
Lieutenant	Lord Douglas Compton		Lord Douglas Compton		G. F. Henry	
,,	H. F. Stanley		H. F. Stanley		G. P. Ellison	
,,	H. E. W. Fiennes		H. E. W. Fiennes		D. G. Campbell	
,,	G. F. Henry		G. F. Henry		Lord C. Bentinck	
,,	G. P. Ellison		G. P. Ellison		W. von Schröder	
,,	D. G. Campbell		D. G. Campbell		D. J. Beale-Browne (Adjutant)	13 Mar. 1895
,,	Lord Chas. Bentinck		Lord C. Bentinck		E. F. Bell	5 Sept. 1894
,,	W. von Schröder		W. von Schröder		J. S. Forbes	7 Nov. 1894
,,	..		D. J. Beale-Browne			
2nd Lieut.	J. W. Russell		Ernest F. Bell		F. H. Allhusen	
,,	D. J. Beale-Browne		James S. Forbes	25 Jan. 1893	T. Malcolm H. Kincaid-Smith	
,,	E. F. Bell		Fred. H. Allhusen	,,	Victor Reginald Brooke	12 Dec. 1894
R.-Master	E. H. Nicholson		E. H. Nicholson		E. H. Nicholson	
Quarter-master	J. Forsythe		D. Laing (Honorary Lieut.)	23 Dec. 1893	David Laing	

NINTH (QUEEN'S ROYAL) LANCERS—*continued.*

Rank.	1896	Date of Commission.	1897	Date of Commission.	1898	Date of Commission.
Colonel	Sir Wm. Drysdale, K.C.B.		Sir Wm. Drysdale, K.C.B.		Sir Wm. Drysdale, K.C.B.	
Lieut.-Col.	Bloomfield Gough	15 Dec. 1895	B. Gough		B. Gough	
Major	M. O. Little		M. O. Little		M. O. Little	
,,	Spencer W. Follett	13 Feb. 1895	S. W. Follett		S. W. Follett	
,,	..		C. Bishop	15 Dec. 1895	C. Bishop	
Captain	C. Bishop					
,,	W. K. Jenner		W. K. Jenner		W. K. Jenner	
,,	F. F. Colvin		F. F. Colvin		F. F. Colvin	
,,	F. W. Duff		The Hon. C. Willoughby		The Hon. C. Willoughby	
,,	The Hon. C. Willoughby		Lord D. Compton		F. T. Lund	
,,	F. T. Lund		H. F. Stanley		Lord D. Compton	
,,	Lord D. Compton		H. E. W. Fiennes		H. F. Stanley	
,,	H. F. Stanley		G. F. Henry		H. E. W. Fiennes	
,,	H. E. W. Fiennes		G. P. Ellison	,,	G. F. Henry	
,,	G. F. Henry	1 July 1895	E. R. Gordon	18 Sept. 1895	G. P. Ellison	
Lieutenant	G. P. Ellison		D. G. Campbell		E. R. Gordon	
,,	D. G. Campbell		Lord C. Bentinck		D. G. Campbell	
,,	Lord C. Bentinck		D. J. Beale-Browne		Lord C. Bentinck	
,,	W. von Schröder		E. F. Bell		D. J. Beale-Browne	
,,	D. J. Beale-Browne		F. H. Allhusen		E. F. Bell (Adjutant)	20 April 1898
,,	E. F. Bell		T. M. Kincaid-Smith		F. H. Allhusen	
,,	J. S. Forbes		V. R. Brooke	29 April 1896	T. M. Kincaid-Smith	
					V. R. Brooke	
,,	F. H. Allhusen	13 Mar. 1895	G. Skeffington-Smyth	4 July 1896	G. Skeffington-Smyth	
,,	T. M. Kincaid-Smith	25 Sept. 1895	J. Gordon-Stirling	,,	J. Gordon-Stirling	
2nd Lieut.	V. R. Brooke		Henry M. Durand	13 May 1896	H. M. Durand	
,,	Geoffrey H. J. Skeffington-Smyth	29 May 1895	William S. Thorpe	5 Aug. 1896	D. Piggott-Carleton	
,,	J. Gordon-Stirling	31 July 1895	Dudley M. Piggott-Carleton	,	Eustace H. Abadie	11 Aug. 1897
,,	..		..		Lord Fred. T. H. Temple-Blackwood	,,
,,	..		..		Fred. W. Cavendish	8 Sept. 1897
,,	..		..		Stanley R. Theobold	,,
,,	..		..		J. Stewart Duckett	1 Dec. 1897
Riding-Master	George Parker	27 Feb. 1895	William Baxter	5 Feb. 1896	W. Baxter	
Qr.-master	D. Laing.		D. Laing		D. Laing	

NINTH (QUEEN'S ROYAL) LANCERS—*continued.*

Rank.	1899	Date of Commission.	1900	Date of Commission.	1901	Date of Commission.
Colonel	Sir W. Drysdale, K.C.B.		Sir W. Drysdale, K.C.B.		H. A. Bushman, C.B. (Maj.-Gen. 8 Aug. 1900)	
Lieut.-Col.	B. Gough		B. Gough		M. O. Little	15 Mar. 1900
Major	M. O. Little		M. O. Little		S. W. Follett	
,,	S. W. Follett		S. W. Follett		C. Bishop	
,,	C. Bishop		C. Bishop		W. K. Jenner	
,,	W. K. Jenner	12 Mar. 1898	W. K. Jenner		F. F. Colvin	,,
Captain	F. F. Colvin		F. F. Colvin		The Hon. C. Willoughby	
,,	The Hon. C. Willoughby		The Hon. C. Willoughby		F. T. Lund	
,,	F. T. Lund		F. T. Lund		Lord D. Compton	
,,	Lord D. Compton		Lord D. Compton		H. E. W. Fiennes	
,,	H. F. Stanley		H. F. Stanley		E. R. Gordon	
,,	H. E. W. Fiennes		H. E. W. Fiennes		D. G. Campbell	
,,	G. F. Henry		G. P. Ellison		Lord C. Bentinck	8 Nov. 1900
,,	G. P. Ellison		E. R. Gordon			
,,	E. R. Gordon		D. G. Campbell	3 May 1899		
Lieutenant	D. G. Campbell		Lord C. Bentinck		D. J. Beale-Browne	
,,	Lord C. Bentinck		D. J. Beale-Browne		E. F. Bell	
,,	D. J. Beale-Browne		E. F. Bell		F. H. Allhusen	
,,	E. F. Bell		F. H. Allhusen		T. M. Kincaid-Smith	
,,	F. H. Allhusen		T. M. Kincaid-Smith		V. R. Brooke	
,,	T. M. Kincaid-Smith		V. R. Brooke		G. Skeffington-Smyth	
,,	V. R. Brooke		G. Skeffington-Smyth		J. Gordon-Stirling	
,,	G. Skeffington-Smyth		J. Gordon-Stirling		H. M. Durand	
,,	J. Gordon-Stirling		H. M. Durand		D. Pigott-Carleton	
,,	H. M. Durand	,,	D. Pigott-Carleton		E. H. Abadie	
,,	D. Pigott-Carleton	2 May 1898	E. H. Abadie	,,	Lord F. Blackwood	9 Oct. 1899
,,	..		..		F. W. Cavendish	8 June 1900
,,	..		..		S. R. Theobald	3 Oct. 1900
,,	..		..		J. S. Duckett	,,
,,	..		..		L. W. S. Jackson	,,
2nd Lieut.	E. H. Abadie		Lord F. Blackwood		G. G. Cobden	
,,	Lord F. Blackwood		F. W. Cavendish		R. V. Wynn	
,,	F. W. Cavendish		S. R. Theobald		A. S. Trower	
,,	S. R. Theobald		J. S. Duckett		Hon. Arch. Ronald Macdonald	21 Feb. 1900
,,	J. S. Duckett		L. W. S. Jackson		Charles, Earl of Leitrim	21 April 1900
,,	Lionel W. Sadleir-Jackson	4 May 1898	G. G. Cobden		Richard S. Chadwick	7 July 1900
,,	George G. Cobden	3 Aug. 1898	John Fred. Pollok	4 Jan. 1899	Douglas K. Lucas Tooth	8 Aug. 1900
,,	..		Percy Fred. Brassey	27 May 1899	George M. Neilson	11 Aug. 1900
,,	..		Robert Vaughan Wynn	28 Oct. 1899		
,,	..		Arthur Seymour Trower	4 Nov. 1899		
Riding-Master	W. Baxter					
Qr.-master	D. Laing					

NINTH (QUEEN'S ROYAL) LANCERS—*continued.*

Rank.	1902	Date of Commission.	1903	Date of Commission.
Colonel	H. A. Bushman, C.B. (Maj.-Gen.)		H. A. Bushman, C.B. (Maj.-Gen.)	
Lieut.-Col.	M. O. Little		M. O. Little (Col. 29 Nov. 1900)	
Major	S. W. Follett		S. W. Follett (Brev. Lt.-Col. 22 Aug. 1902)	
,,	W. K. Jenner		F. F. Colvin (Brev. Lt.-Col. 29 Nov. 1900)	
,,	F. F. Colvin		The Hon. C. Willoughby	
,,	The Hon. C. Willoughby	9 Jan. 1901	F. T. Lund (Brev. Lt.-Col. 26 June 1902)	27 Sept. 1902
Captain	F. T. Lund		Lord D. Compton (Brev. Maj. 5 Sept. 1901)	
,,	Lord D. Compton		E. R. Gordon	
,,	H. E. W. Fiennes		D. G. Campbell (Bt.-Maj. 22 Aug. 1902, Adj. 6 Sept. 1902)	
,,	E. R. Gordon		Lord C. Bentinck (Bt.-Maj. 29 Nov. 1900)	
,,	D. G. Campbell		D. J. Beale-Browne	
,,	Lord C. Bentinck		E. F. Bell	
,,	D. J. Beale-Browne	1 Jan. 1901	V. R. Brooke, D.S.O.	
,,	E. F. Bell	,,	G. Skeffington-Smyth	
,,	F. H. Allhusen			
,,	V. R. Brooke	6 May 1901		
,,	G. Skeffington-Smyth	3 June 1901		
Lieutenant	T. M. Kincaid-Smith		T. M. Kincaid-Smith	
,,	J. Gordon-Stirling (Adj. 20 April 1902)		H. M. Durand	
,,	H. M. Durand		D. Pigott-Carleton	
,,	D. Pigott-Carleton		E. H. Abadie, D.S.O.	
,,	E. H. Abadie		Lord Frederick Blackwood, D.S.O.	
,,	Lord F. Blackwood		F. W. Cavendish	
,,	F. W. Cavendish		J. S. Duckett	
,,	J. S. Duckett		L. W. S. Jackson, D.S.O.	
,,	L. W. S. Jackson		G. G. Cobden	
,,	G. G. Cobden	1 Jan. 1901	R. V. Wynn	
,,	R. V. Wynn	6 May 1901	A. S. Trower	
,,	A. S. Trower	3 June 1901	R. S. Chadwick	13 Aug. 1901
,,	Lord Leitrim		D. K. Tooth	,,
2nd Lieut.	R. S. Chadwick		G. M. Neilson	
,,	D. K. Tooth		R. H. Brocklebank	
,,	George M. Neilson		H. F. Wood	
,,	Richard H. Brocklebank	5 Jan. 1901	George, Earl of Rocksavage	
,,	Herbert F. Wood	8 May 1901	The Hon. R. B. Cole	
,,	George, Earl of Rocksavage	19 Oct. 1901	K. C. Zarzhetsky	
,,	Hon. R. B. Cole	,,	H. A. Chissendale-Marsh	11 Sept. 1903
,,	Karl C. Zarzhetsky	18 Jan. 1902	A. N. Edwards	August 1903
Riding-Master	..		John H. Kemple	25 Oct. 1902
Qr.-master	..		D. Laing	

APPENDIX VIII.

ALPHABETICAL LIST OF OFFICERS GAZETTED TO THE REGIMENT.

Name.	Appointed to 9th Lancers.		Remarks.
	Date.	Rank.	
Abadie, Henry R. . . .	August 1873	Captain	From half pay.
Abadie, Eustace H. . .	August 1897	2nd Lieut.	
Adams, Christopher . .	July 1715	Cornet.	
Adams, William . . .	May 1808	Surgeon	
Adams, J. P.	Dec. 1874	Vet.-Surgeon	
Adlercron, John . . .	Nov. 1757	Cornet	Resigned September 1761.
Adlercron, William . . .	April 1762	,,	
Agnew, Charles . . .	June 1858	,,	Exchanged to 16th Lancers.
Aldworth, St. Leger . .	1741	,,	
Allan, Andrew . . .	May 1842	Quartermaster	Private, Royal Horse Guards, November 1824.
Allen, Charles	Sept. 1812	Cornet	
Allen, William. . . .	Oct. 1851	Major	Exchanged from 1st Dragoon Guards with Major Spottiswoode. In 1854 exchanged to 16th Lancers, with Major Foster.
Allhusen, Frederick H. . .	Jan. 1893	2nd Lieut.	
Alridge, Robert . . .	May 1800	Captain	
Alston, Harry R. . . .	Jan. 1880	2nd Lieut.	From 18th Hussars.
Amherst, ——	1733	Cornet	Died 1735.
Amyud, Theodore . . .	June 1720	Surgeon	
Anderton, W. Francis . .	May 1841	Cornet	Died March 1847.
Anfrere, George . . .	Sept. 1816	Lieut.	
Anson, O. H. St. George .	April 1842	,,	Transferred from 3rd Foot. Died 1859
Antrobus, Philip . . .	April 1842	Cornet	Exchanged to 91st Foot with Lieut. Head, 1847.
Apperley, Herbert W. . .	Dec. 1875	Captain	Exchanged from 11th Hussars with Captain Grissell.
Arbuthnot, W. W. . . .	Dec. 1848	Cornet	Promoted 63rd Foot.
Archer	1724	Surgeon	
Arkwright, Eustace . .	Jan. 1840	Cornet	Exchanged to 4th Hussars with Lieut. Kemp.
Armstrong, Anthony . .	Feb. 1773	Chaplain	
Armstrong, Alfred F. . .	March 1787	Cornet	
Armstrong, W. Bloss . .	March 1812	,,	
Aston, ——	1727	,,	
Atkinson, George . . .	August 1848	Lieut.	Exchanged from 89th Foot.
Aylesford, Heneage, Earl of .	May 1812	,,	
Babington, Thomas . .	Feb. 1738	,,	
Bacon, John William . .	Jan. 1813	,,	
Bacon, Charles . . .	June 1817	,,	
Bagwell, Hamilton . . .	May 1794	Cornet	
Bailey, Frederick . . .	March 1858	Vet.-Surgeon	From 1st Dragoon Guards. Retired 1861.
Baldwin, John. . . .	May 1800	Cornet	
Barlow, Cuthbert . . .	July 1848	Paymaster	By exchange with Ratcliffe.
Barlow, Sir Morison, Bart. .	1870	Captain	

Name.	Appointed to 9th Lancers.		Remarks.
	Date.	Rank.	
Barnes, Bartholemew	Sept. 1746	Cornet	
Barras, William	Sept. 1811	Lieut.	
Barrow, Leslie A.	July 1865	Cornet	
Barrymore, Richard, Earl of	Oct. 1766	Captain	Aide-de-Camp to the Lord-Lieut. of Ireland, 1767. Resigned April 1769.
Barton, Robert J.	Sept. 1866	Cornet	Exchanged to the Coldstream Guards with Captain the Hon. H. Legge, 1873. Killed in action, Inhlobana, March 1879.
Bashford, C. Browne	March 1858	,,	
Bassano, C. B.	Feb. 1855	Assist.-Surgeon	Never joined.
Baxter, William	Feb. 1896	Riding-Master	Died at Sialkote, 15 August 1902. (Enteric fever.)
Bayley, Samuel	May 1799	Cornet	
Beale-Browne, D. J.	May 1891	2nd Lieut.	
Beasley, Richard	June 1812	Cornet	
Beaver or Beevor, John	March 1772	,,	
Bell, Ernest F.	Dec. 1891	2nd Lieut.	
Bentinck, Lord George	Feb. 1819	Cornet	
Bentinck, Lord Charles	June 1899	2nd Lieut.	
Beresford, William	Dec. 1819	Cornet	
Beresford, Lord William	June 1867	,,	
Bernard, The Hon. Francis	June 1808	,,	
Bertie, The Hon. Charles B.	Oct. 1839	,,	To 2nd Life Guards, 1840.
Bingham, Sir Charles, Bart.	April 1750	,,	Created Baron of Lucan, 1776.
Bird, William W.	Nov. 1843	,,	
Bishop, Charles	August 1879	2nd Lieut.	
Blackwood, Lord Frederick T.	August 1897	,,	
Blair, Robert	Dec. 1853	Cornet	From the 16th Lancers. Exchanged to 2nd Dragoon Guards with Lieut. Chadwick.
Blake, John	May 1767	,,	
Blake, Matthew	Feb. 1804	Captain	
Blake, William G.	Feb. 1861	Lieut.	Died, 25th July 1904.
Blakeney, ——	1733	Lieut.-Colonel	
Blakiston, Charles	June 1820	Captain	
Blennerhasset, Arthur	June 1742	Lieut.	An M.P. December 1745.
Bolton, Loftus	March 1719	Cornet	
Bontein, Edward T.	Dec. 1809	Captain	
Boss, George	June 1809	Cornet	
Brassey, Percy F.	May 1899	2nd Lieut.	
Breslin, W. J.	Nov. 1828	Assist.-Surgeon	Exchanged from the 63rd Regiment.
Brewer, John	1772	Cornet	
Brickman, Christopher	March 1852	,,	Did not join the "Service" Troops.
Brinkley, John	Feb. 1850	,,	Died 1851.
Briscoe, Horton C.	Sept. 1806	Captain	
Broadhead, B.	May 1832	Lieut.	By exchange from 80th Regiment with Lieut. R. E. Fullerton.
Brocklebank, Richard H.	Jan. 1901	2nd Lieut.	
Brooke, C. Clement	June 1858	Riding-Master	To Royal Military College, Sandhurst.
Brooke, Victor R.	Dec. 1894	2nd Lieut.	
Broughton, Ralph	April 1883	Lieut.	Killed at Simla, 17 July 1885, by fall from his pony.
Brown, John	May 1742	Colonel Comdg.	
Brown, James	Jan. 1752	Cornet	
Brown, Thomas	May 1766	,,	
Browne, Marcus	August 1765	,,	Resigned 1766.
Browne, John	April 1782	,,	Promoted Lieut. 14th Light Dragoons, 1790.
Browne, Benjamin C.	June 1813	,,	
Browne, Lord J. de Burgh	July 1843	,,	Promoted 10th Hussars, 1846.
Browne, J. W. Hamilton	Sept. 1865	,,	
Bulkeley, Robert S. W.	Dec. 1859	Captain	Exchanged from 7th Hussars with Captain Wilkinson.
Burroughs, Thomas	May 1719	,,	
Burton, Edward	July 1818	Assist.-Surgeon	

Name.	Appointed to 9th Lancers.		Remarks.
	Date.	Rank.	
Bury, Robert	March 1858	Cornet	Exchanged to 7th Dragoon Guards.
Bush (*e*), Paul	Sept. 1751	,,	
Bushman, Henry A. . .	Nov. 1858	,,	To 7th Hussars as Cornet, 1859. Returned to 9th Lancers as Captain, March 1872.
Butler, Lord James H. . .	May 1866	,,	Died 1867.
Butson, S. G.	July 1870	,,	Killed in action, 13 Dec. 1879.
Butterfield, Thomas .	Oct. 1798	Adjutant	
Caldwell, Josiah . . .	Dec. 1784	Cornet	
Caldwell, John . . .	Dec. 1785	,,	
Call, George Isaac . . .	August 1811	Lieut.	
Callaghan, George . . .	Dec. 1827	Captain	
Cameron, Cyril St. C. . .	Jan. 1880	2nd Lieut.	Transferred from K. D. Guards.
Campbell, Alexander . .	July 1813	Captain	Died 1850.
Campbell, J. D. L. . . .	July 1831	Cornet	Transferred from 16th Lancers, vice Trueman.
Campbell, J. Cameron . .	June 1841	Lieut.	By exchange with Lieut. Sutherland.
Campbell, William R. . .	April 1842	,,	On augmentation. Exchanged to 5th Dragoon Guards with Captain Johnson, 1853.
Campbell, David G. . .	March 1889	2nd Lieut.	
Candy, Henry A. . . .	April 1864	Cornet	
Carden, John Craven . .	March 1774	,,	
Carew, J. Hallowell . .	June 1841	,,	Promoted into 16th Lancers, 1843.
Carleton, William . . .	July 1715	,,	Died. January to March 1740.
Carleton, William Napier .	Feb. 1858	,,	Transferred from 13th Light Dragoons. To 96th Foot, 1861.
Carleton-Pigott, Dudley M. .	August 1896	2nd Lieut.	
Carmichael, Frederick . .	Oct. 1803	Lieut.	
Carnegie, John . . .	Jan. 1832	Captain	From 2nd R. N. B. Dragoons, vice Captain Fleetwood-Shawe, to half-pay.
Carter, Arthur	Jan. 1777	Lieut.	
Catheray, W.	Never	joined.	To 6th Dragoons.
Caulfield, Toby . . .	June 1771	Lieut.	An M.P. 1772.
Cave-Browne-Cave, B. W. .	Jan. 1861	Cornet	
Cavendish, H. F. C. . .	Sept. 1818	Major	
Cavendish, Frederick W. .	Sept. 1897	2nd Lieut.	
Chabot, William . . .	March 1809	Major	Lieut.-Colonel in the army.
Chadwick, J. De Heley . .	Jan. 1857	Lieut.	Exchanged from 2nd Dragoon Guards with Lieut. Blair.
Chadwick, Richard S. . .	July 1900	2nd Lieut.	
Chambers, Richard . . .	Feb. 1875	Captain	Exchanged from 5th Lancers with Captain Green.
Chapman, Benjamin . .	July 1800	Cornet	
Chaytor, John	April 1861	,,	
Cheeseman, E. J. . . .	July 1870	Vet.-Surgeon	
Chisholme-Scott, J. J. . .	Jan. 1872	Sub-Lieut.	To 5th Lancers. Exchanged with Major Morland. Killed when in command of the Imperial Light Horse at Elandslaagte, South Africa, 21 Oct. 1899.
Clarke, Henry	1738	Cornet	
Clarke, P. Haughton . .	May 1837	,,	
Clarke, Perceval . . .	Feb. 1861	,,	
Clayton, Edward . . .	1727	,,	Resigned August 1760.
Clayton, William C. . .	June 1867	Captain	By exchange. Killed at polo, Delhi, 25 Dec. 1876.
Cleland, Robert S. . . .	Nov. 1860	Lieut.	From 7th Dragoon Guards. Died 7 Aug. 1880, from wounds received in action 11 Dec. 1879.
Clerk, James	Dec. 1833	,,	Exchange with Lieut. Coghlan.
Clerke, John	May 1805	,,	
Clery, George C. . . .	Oct. 1860	Assist.-Surgeon	By exchange with Assistant-Surgeon Fuller.
Clifford, John	Jan. 1854	,,	Exchange with Assistant-Surgeon Jephson.

Name.	Appointed to 9th Lancers.		Remarks.
	Date.	Rank.	
Clifton, Chandos F.	May 1844	Cornet	Promoted into 12th Lancers, 1850.
Cobden, George Gough	August 1898	2nd Lieut.	
Coddington, Dixie	June 1753	Cornet	Resigned. April to June 1766.
Coen, ——	August 1779	,,	
Coghlan, Roger	July 1833	Lieut.	Exchanged from 3rd L. Dragoons, then exchanged to 83rd Regiment with Lieut. Clerk.
Coldclough, A. Cæsar	Jan. 1822	Surgeon	By exchange with Surg. Melin.
Cole, The Hon. Arthur	Sept. 1772	Cornet	
Cole, Henry	Jan. 1788	,,	
Cole, The Hon. Reginald B.	Oct. 1901	2nd Lieut.	
Coles, Josias R. J.	Dec. 1851	Captain	From 4th Light Dragoons on exchange with Captain Donovan.
Colthurst, James	May 1777	Ensign	
Colville, Charles J.	Never	joined	Promoted Canadian Rifles, 1844.
Colvin, Forrester F.	Dec. 1880	2nd Lieut.	
Compton, Lord Douglas	Sept. 1885	,,	
Cooke, Robert	April 1827	Cornet and Adjutant	Private, 9th Lancers, from Oct. 1804.
Cooper, The Hon. A. H.	May 1841	Captain	Never joined.
Cope, Sir John	June 1737	Colonel Comdg.	
Corbet, Andrew	Nov. 1780	Captain	
Cornwall, John F.	August 1799	Cornet	
Cosby, Thomas	June 1800	,,	
Costwick, Sir William, Bart.	1724	,,	
Cox, Thomas	Feb. 1716	,,	Promoted from Quartermaster.
Craiggs, ——	May 1720	,,	
Crawford, Henry	July 1715	Lieut. and Captain	At this date and later the senior Lieut. had the rank of Captain, and commanded the Colonel's troop.
Crawford, Alexander S.	May 1801	Lieut.	
Crofton, Thomas	Dec. 1738	Cornet	Resigned 1749.
Crofton, Frederick R.	Feb. 1858	,,	Exchanged to 7th Dragoon Guards, 1860.
Crofts, James	July 1719	Colonel Comdg.	
Crowdy, John	Nov. 1866	Riding-Master	Private, 14th Lt. Dragoons, June 1855. To 5th Lancers, 1875.
Cullen, John	1741	Cornet	
Curteis, Edward B.	Jan. 1861	Lieut.	
Curtis, Thomas	Oct. 1769	Cornet	Resigned Sept. 1774.
Dallway, ——	Oct. 1729	Lieut.-Colonel	
Dalton, Philip Tuite	Oct. 1811	Cornet	
Daly, Hyacinth	Sept. 1786	,,	
Daly, Dennis	Oct. 1811	Captain	
Darby, Thomas	July 1737	Cornet	
Darvall, Edward	Nov. 1799	Captain	
Daulhat, Christopher	1747	Cornet	
Davies, T. C.	Oct. 1804	Assist.-Surgeon	
Dawson, Henry	April 1799	Cornet	
Dawson-Massey, C.	April 1842	Captain	Exchanged from 5th Foot with Captain Isacke.
Deacon, Charles	March 1846	,,	Exchanged from 13th Hussars with Captain Nicholson.
D'Esté, Augustus F.	April 1817	,,	
Degennes, J. Daniel	Sept. 1739	Lieut.-Colonel	
Delancey, James	May 1808	,,	
Delmar, C. Abbott	May 1846	Cornet	
Denis, Humphrey	Dec. 1784	Captain	
Desbrisay, Thomas	Sept. 1747	Cornet	
Dick, Charles	Sept. 1806	,,	
Digby, E. St. Vincent	July 1828	Lieut.	Exchanged from 65th Regiment. Succeeded as Lord Digby, 1856.
Dixon, Robert S.	Jan. 1842	Cornet	
Donaghoe, James	May 1797	Assist.-Surgeon	
Donaldson, Richard	July 1774	Cornet	
Donovan, Charles H.	Oct. 1841	Lieut.	Exchanged from 13th Hussars with Lieut. Madocks. Exchanged to 4th Light Dragoons with Captain Coles, 1851.

Name.	Appointed to 9th Lancers.		Remarks.
	Date.	Rank.	
Douglas, William	May 1720	Captain	
Dressner, Charles J.	Sept. 1876	Sub-Lieut.	
Drysdale, William	April 1842	Lieut.	From 4th Light Dragoons.
Duckett, J. Steuart	Dec. 1897	2nd Lieut.	
Duff, Frederick W.	Feb. 1881	,,	
Dunbar, John	1715	Major	Died 1724.
Duncombe, William	July 1715	Lieut.-Colonel	
Dunn, John W.	July 1811	Lieut.	
Durand, Henry	May 1896	2nd Lieut.	
Ebhart, William	July 1803	Lieut.	
Ebhart, Frederick	August 1805	,,	
Edwards, Henry W.	Nov. 1863	Cornet	
Edwards, A. N.	August 1903	2nd Lieut.	
Ellice, Robert	Feb. 1800	Lieut.	
Ellice, George	July 1802	Cornet	
Elliot, Alexander J.	July 1848	,,	To 5th Dragoon Guards, 1850.
Ellis, Edward	June 1741	,,	
Ellis, Augustus F.	April 1817	,,	
Ellis, William S. A.	Dec. 1836	Lieut.	Transfer from 16th Lancers, vice Lieut. Ibbetson.
Ellis, Frederick	August 1847	Cornet	
Ellis, Charles J. H.	April 1852	,,	To 6th Dragoon Guards, 1856.
Ellison, ——	May 1794	,,	
Ellison, George P.	August 1888	Sub-Lieut.	Died South Africa.
Erskine, John Francis	July 1757	Cornet	
Erskine, The Hon. William J.	Oct. 1794	Captain	Killed during the Irish Rebellion, 24 May 1798.
Erskine, William M.	May 1860	Lieut.	Exchanged from 7th Dragoon Guards with Lieut. Crofton. Succeeded as 5th Lord Erskine, 1882.
Evans, John	April 1856	Cornet	Died 1903. Was in the 32nd Regiment.
Evans, George	Feb. 1858	Assist.-Surgeon	
Evans, G. A. P.	June 1874	Sub-Lieut.	Exchanged to 7th Hussars with Captain Reynard, 1886.
Fawcett, Anthony M.	Dec. 1855	Captain	Exchanged from 50th Regiment with Captain F. C. Trower. To 2nd Dragoon Guards in 1866.
Fenton, Michael	July 1784	Cornet	
Ferguson, Samuel	Oct. 1797	Vet.-Surgeon	
Ferguson, Henry	March 1818	Cornet	
Ferrers, Lord George	Sept. 177c	,,	
Fiennes, The Hon. Ivo-de-Vesci	June 1864	Major	Exchanged from 7th Hussars. Died 23 November 1875.
Fiennes, Henry E. J. Wykeham	April 1887	2nd Lieut.	
Fife, William H.	May 1868	Cornet	
Fincastle, Alexr., The Viscount	August 1826	,,	Exchanged to 3rd L. Dragoons with Lieut. Spalding, 1827.
Fitzgerald, Richard	June 1739	Lieut.	
Fitzmaurice, The Hon. W. E.	Nov. 1824	Cornet	
Fleming, John L.	Oct. 1795	,,	
Floyd, Charles A.	July 1859	,,	
Follett, Spencer W.	Feb. 1895	Major	From 4th Hussars.
Folliot, Lewis	July 1715	Cornet	
Folliot, John	Jan. 1716	,,	
Forbes, James	Jan. 1893	2nd Lieut.	
Forster, Thomas	Oct. 1806	Lieut.	
Forsythe, John	Sept. 1879	Quartermaster	Private 9th Lancers Oct. 1858. Retired 23 Oct. 1893.
Fortescue, John	June 1785	Chaplain	
Foster, Charles J.	March 1854 Never	Major joined	
Francis, Thomas J.	Sept. 1842.	Lieut.	Exchanged from 4th L. Dragoons with Lieut. Magan. Promoted 16th Lancers, 1855.
Franklin, Henry	April 1856	Surgeon	Vice Clifford.

Name.	Appointed to 9th Lancers. Date.	Appointed to 9th Lancers. Rank.	Remarks.
French, Robert	Nov. 1746	Cornet	
French, James Boyle	August 1811	,,	
French, Lucius Henry	April 1842	Captain	From 1813 to 1824 was in the Royal Navy. In 1826 in the 11th L. Dragoons. Died Dec. 1842, on the march from Calcutta to Cawnpore.
French, Lucius John	Feb. 1843	Cornet	From 31st Foot. Killed in action at Agra, 1857.
Fuller, Samuel	Sept. 1857	Assist.-Surgeon	Vice Clery.
Fuller, George	Oct. 1869	Quartermaster	Private, 9th Lancers, Aug. 1843. Died from sickness, Afghanistan, 1879.
Fullerton, J. A.	August 1822	Cornet	Died in Cashmere, 28 April 1850.
Fullerton, R. E.	March 1831	,,	To 80th Regiment on exchange with Lieut. Broadhead.
Gabbott, Richard	July 1777	Ensign	
Gain, Joseph	March 1808	Vet.-Surgeon	
Gale, William	Dec. 1790	Lieut.	
Garden, James	Dec. 1764	Cornet	Was a Quartermaster 1751 or earlier.
Gaskell, Henry B.	Oct. 1865	,,	
Gerard, Frederick	July 1831	,,	
Gifford, Walter B.	Oct. 1855	,,	Promoted to 6th Dragoon Guards, 1856.
Gilbert, ——	1725	,,	
Gillam, Isaac	Sept. 1812	Lieut.	
Goddard, Henry	April 1724	Major	Retired 1729.
Goldie, Thomas	May 1799	,,	
Goldie, James	July 1854	Cornet	Exchanged to 17th L. Dragoons with Captain Marshall, 1860.
Gordon, William C.	Feb. 1852	,,	
Gordon, E. R.	Sept. 1895	Captain	From 7th Dragoon Guards. Transferred.
Gore, Arthur	May 1806	Lieut.	
Gore, George	Sept. 1806	Captain	
Goldsmid, J. Louis	Nov. 1808	Cornet	
Goldsworthy, Roger T.	June 1865	Lieut.	
Gorham, Richard	March 1800	,,	
Gough, Maurice	April 1758	Chaplain	
Gough, Bloomfield	April 1873	Lieut.	From Rifle Brigade.
Grainger, R. J.	July 1856	Cornet	Never joined.
Grangue, Henry de	April 1743	Colonel Comdg.	
Grant, C.	Sept. 1800	Captain	
Grant, Montagu H.	April 1816	Cornet	
Grant, Hope-, James	August 1826	,,	
Grant, Robert J. G.	Dec. 1846	Surgeon	Vice Wood.
Grant, Francis R. C.	Sept. 1851	Cornet	Promoted 5th L. Dragoons, 1858.
Grant, Maximilian	Jan. 1867	Assist.-Surgeon	
Grant, W.	22 Dec. 1903	Quartermaster	Was Regt.-Sergeant-Major of the Regiment.
Grattan, James	July 1811	Lieut.	
Green, William	June 1720	Cornet	
Green, Philip	Dec. 1866	,,	To 5th Lancers on exchange with Captain Chambers, 1875.
Greenwood, Julius	Sept. 1816	,,	
Griffiths, Frederick C.	May 1855	Lieut.-Colonel	Never joined.
Grissell, F. De la G.	March 1865	Cornet	Exchanged to 11th Hussars with Captain Apperley, 1875.
Gubbings, Lancelot	Dec. 1792	,,	
Guernsey, Lord	August 1811	,,	See Aylesford.
Gurdon, ——	May 1814	,,	
Gurley, William	April 1805	Lieut.	
Gurwood, John	July 1812	Captain	
Hall, James	June 1782	Cornet	
Hall, Henry J.	July 1858	,,	
Hames, John	May 1805	,,	
Hamilton, Gustavus	1715	Lieut.	
Hamilton, William	April 1842	Cornet and Adjutant	Private, 9th Lancers, Feb. 1833. Died 1858 at Calcutta.

Name.	Appointed to 9th Lancers.		Remarks.
	Date.	Rank.	
Handley, Benjamin . .	Nov. 1805	Lieut.	
Handley, Benjamin . .	May 1809	Cornet	
Hankey, William . . .	Jan. 1828	"	
Hardy, John	March 1862	Captain	From 6th Dragoons.
Harley, Lord Alfred . .	Never	joined	
Harrison, Thomas . . .	April 1825	Vet.-Surgeon	
Hawley, Sir Joseph, Bart. .	Aug. 1832	Cornet	
Hawtry or Hawtrey, Alexander	April 1842	"	
Hay, Lord Charles . . .	1729	Captain	Promoted to another regiment, 1742.
Hay, Thomas	June 1751	Cornet	
Hayes, John	Sept. 1804	"	
Hayne, George W. . . .	Sept. 1874	Captain	
Head, John	June 1847	Lieut.	Exchanged with Lieut. Antrobus.
Hearsey, Charles J. R. .	Feb. 1875	Sub-Lieut.	Killed in action, 11 Dec. 1879.
Hely, Thomas	Nov. 1809	Quartermaster	
Henniker, B. Trecothick .	March 1793	Lieut.-Colonel	
Henry, Francis . . .	June 1863	Cornet	
Henry, Gilbert F. . . .	Nov. 1887	2nd Lieut.	
Hepburn, John B. . . .	March 1860	Cornet	
Herbert, Francis A. . .	July 1866	"	
Herries, William L. . .	April 1804	Lieut.	
Hewitt, The Hon. Wm. Henry	June 1770	Cornet	
Heydon, John	April 1841	Quartermaster	Private, 9th Lancers, Sept. 1807. Exchanged to 3rd Dragoon Guards with Quartermaster Allen, 1842.
Higgins, J. Lewis . . .	March 1795	Lieut.	
Hill, James	July 1715	Cornet	
Hippersley, Gustavus . .	Dec. 1787	"	
Hives, Alfred	Oct. 1849	"	
Hoghton, Richard . . .	Sept. 1846	"	
Honywood, Philip . . .	May 1756	Colonel Commanding	
Hopkins, Henry C. . . .	August 1876	Lieut.	
Horsburgh, Thomas . .	August 1805	Cornet	
House, Peter	Nov. 1852	Quartermaster	Private, 9th Lancers, Oct. 1840. Appointed Quartermaster Cavalry Depot, 1861.
Howard, Charles . . .	May 1719	Captain	
Howard, The Hon. Charles F. A.	June 1866	Lieut.	
Hughes, Henry . . .	June 1753	Cornet	
Hull, William	March 1761	Captain	Promoted to another regiment, 1762.
Humberston, James A. . .	1867	Cornet	
Humbley, Wm. W. Waterloo .	April 1842	Lieut.	On augmentation. Exchanged to 26th Foot, 1856.
Humphreys, William . .	July 1715	"	A Captain in 1737.
Humphries, —— . . .	April 1804	Surgeon	
Hunt, William . . .	May 1783	Captain	
Hunt, Edmund . . .	Oct. 1847	Cornet	Promoted 6th Dragoons, 1850.
Hunter, James . . .	March 1878	2nd Lieut.	
Huntley, W. W. . . .	Oct. 1817	Lieut.	From 1st Dragoon Guards. Left in 1820; returned in 1841 as a Captain. Died 1843.
Hurford, R. Gedaliah . .	August 1844	Vet.-Surgeon	Exchanged to 3rd Dragoon Guards with Vet.-Surgeon Shaw, 1857.
Hurt, James	Dec. 1810	Captain	
Hutcheon, ——	Oct. 1804	Cornet and Adjutant	
Hutchinson, Thomas W. .	Jan. 1856	Captain	Exchanged with Capt. Humbley from 26th Foot.
Hyder, W. A.	Never	joined	To 16th Light Dragoons.
Ibbetson, Charles H. . .	April 1833	Cornet	Retired 1836.
Ingham, Charles T. . .	August 1840	Surgeon	Exchanged to 78th Highlanders with Surgeon Wood.
Innes, Anthony . . .	July 1777	Ensign	
Irving, J. C.	1885	Paymaster	

Name.	Appointed to 9th Lancers.		Remarks.
	Date.	Rank.	
Isacke, Frederick J.	Sept. 1836	Cornet	Exchanged to 5th Foot, with Capt. Dawson, 1842.
Jackson, T. Henry	Feb. 1807	,,	
Jackson-Sadleir, Lionel	May 1898	2nd Lieut.	
James, Demetrius	1736	Cornet	
Jenner, Sir Walter K., Bart.	Oct. 1880	2nd Lieut.	
Jenney, Arthur H.	April 1840	Cornet	Exchanged with Lieut. Williams to 64th Foot.
Jephson, William H.	Nov. 1850	Assist.-Surgeon	Exchanged with Assist.-Surgeon Clifford, 1853.
Jevers, William	Nov. 1790	Cornet	
Johnson, C. Philip	May 1853	Captain	Exchanged with Capt. Campbell from 5th Dragoon Guards.
Johnson, Edward P.	July 1879	2nd Lieut.	
Johnston, James	Jan. 1771	Colonel Commanding	
Johnston, Richard	July 1777	Adjutant	
Johnston, Richard	Dec. 1780	Cornet	
Johnston, James	July 1833	,,	
Johnston, George	April 1842	Vet.-Surgeon	Exchanged from 7th Hussars with Vet.-Surgeon Robertson.
Johnston, Hamilton H.	April 1864	Cornet	
Jones, Benjamin	Oct. 1800	,,	
Jones, Charles Tyrwhitt	June 1818	,,	
Jones, Alfred S.	July 1852	,,	Promoted 18th Light Dragoons, 1858.
Jordan, Charles	March 1805	Surgeon	
Jorden, John	April 1756	Colonel Commanding	
Kearnay, James	May 1794	Lieut.	
Keene, Benjamin	June 1808	Captain	
Kemp, Phillip	May 1842	Lieut.	Exchanged with Lieut. Arkwright from 4th Light Dragoons.
Kemple, John H.	Oct. 1902	Riding-Master	
Kerr, Mark	Oct. 1784	Captain	
Kerr, John B.	June 1809	Cornet	
Kidd, Robert C.	July 1850	,,	
Kindillon, Robert	March 1775	,,	
King, Robert W.	June 1843	,,	
King, John Henry	Jan. 1846	,,	Exchanged with Lieut. Upton to 1st West India Regiment, 1855.
King, Edward	Feb. 1856	,,	Never joined.
Kirkwall, John, The Viscount	Sept. 1823	,,	
Knight, Henry	Dec. 1819	Paymaster	
Knightley, M. C.	Oct. 1809	Cornet	
Knott, John	Oct. 1806	Cornet and Adjutant	
Knox, Andrew	July 1715	Captain	
Knox, Mitchelburne	Dec. 1779	Cornet	
Knox, James	August 1814	Assist.-Surgeon	
Knox, George	July 1823	Cornet	
Korke, ——	March 1813	Vet.-Surgeon	
Kortwright, W. Cæsvelt	April 1842	Cornet	
Laing, David	1895	Quartermaster	
Lalor, Thomas	August 1800	Cornet	
Lamont, John Henry	Dec. 1877	Lieut.	
Landers, George	July 1802	Vet.-Surgeon	Killed, Buenos Ayres, 1807.
Langley, John	Nov. 1772	Cornet	
Lascelles, The Hon. Frederick	July 1821	,,	
Lascelles, The Hon. Charles G.	Nov. 1874	Lieut.	Transferred from Royal Horse Guards.
Laski, J. L. de	Sept. 1887	2nd Lieut.	Died at Dundalk, 8 December 1890, from enteric fever.
Latham, William	Jan. 1750	Cornet	
Lauder, Lancelot	April 1794	,,	
Law, Charles E.	Feb. 1844	,,	Promoted 31st Foot, 1846.

Name.	Appointed to 9th Lancers.		Remarks.
	Date.	Rank.	
Lawrell, Walter G.	Dec. 1865	Cornet	To 4th Hussars. Killed in action, Sekukuni's Kraal, Nov. 1879.
Lawrence, Wm. H.	April 1860	,,	Killed, steeplechasing, 1865.
Lawson, Samuel	August 1760	,,	
Lee (Leigh), Edward . .	1735	,,	Vice Amherst deceased. He died 1738.
Leech, Francis	Jan. 1844	Paymaster	
Lefroy, Anthony	June 1785	Lieut.-Colonel	
Legard, Henry W.	Feb. 1828	Cornet	
Legge, The Hon. Heneage .	Dec. 1873	Captain	Exchanged from Coldstream Guards with Captain Barton.
Leitrim, Charles, Earl of .	April 1900	2nd Lieut.	
Lennox, Lord George . .	Feb. 1812	Cornet	
Leslie, John, Lord	July 1715	Captain	Earl of Rothes.
Leslie, A. M.	1889	Paymaster	
Lewer, Robert	Dec. 1871	Surgeon	
Ligonier, Francis	1724	Captain	
Little, Archibald	Oct. 1831	Cornet	
Little, Malcolm O.	Oct. 1878	2nd Lieut.	Transferred from 17th Lancers.
Lodge, Francis	Sept. 1761	Cornet	
Lodge, Barton	March 1768	,,	
Loftus, Edward	Nov. 1749	,,	
Loftus, Henry	Sept. 1790	,,	
Longman, George	Nov. 1861	Vet.-Surgeon	
Lord, William C.	1869	,,	
Loughborough, James, Lord .	Feb. 1819	Cornet	
Love, Robert John	Dec. 1790	,,	
Love, Richard	June 1796	,,	
Luby, ——	Oct. 1806	Assist.-Surgeon	
Lucas, Charles	Feb. 1807	Cornet	
Lukis, Thomas	Feb. 1813	Assist.-Surgeon	
Lund, Frederick T.	May 1892	Captain	Transferred from 3rd Hussars. Commissioned from 12th Lancers.
Lushington, Thomas . .	April 1749	Captain	
Lynch, George	July 1795	Cornet	
Lyster, Mathew	April 1759	,,	
Maberly, William Leader .	July 1817	Lieut.	
Macartney, J. Nixon . .	May 1831	Cornet	Exchanged to 16th Lancers with Captain Pratt, 1847.
Macdonald, The Hon. Archibald	Feb. 1900	2nd Lieut.	
Mackenzie, Lewis	Sept. 1804	Cornet	
Mackenzie, Edward Phillippe	March 1862	,,	
Mackenzie, J. A. Stewart .	May 1867	,,	
Mackintosh, Ronald . .	August 1864	,,	
Macpherson, Eneas M. . .	Feb. 1847	Assist.-Surgeon	
Macrae, Alexander . . .	August 1857	,,	
Madden, Thomas	August 1751	Chaplain	
Madocks, John Edward . .	August 1838	Cornet	Exchanged to 13th Light Dragoons with Lieut. Donovan.
Magan, W. H.	Feb. 1841	,,	Exchanged to 4th Light Dragoons with Lieut. Francis.
Mahon, Bartholemew . .	July 1762	,,	Resigned June 1768.
Mahon, The Hon. Thomas .	Jan. 1797	Lieut.-Colonel	
Mahon, Denis	Sept. 1806	Cornet	
Mahon, Maurice H. . .	March 1859	Paymaster	Exchanged with Paymaster Smales.
Malcolm, John	April 1871	Assist.-Surgeon	
Malide, Paul	1724	Captain	Died December 1748.
Mallett, Hugh	Feb. 1851	Lieut.	
Mallory, Henry	Nov. 1819	,,	
Mann, Henry	March 1754	Adjutant	
Mann, Thomas Henry . .	Jan. 1768	,,	
Mann, Thomas Henry . .	May 1775	Cornet	
Manners, George	Never	joined	
Mansell, The Hon. W. Villiers	May 1807	Lieut.	
Markham, Robert	Dec. 1823	,,	
Marley, George	April 1762	Cornet	
Marsh-Chisendale, H. A. .	Sept. 1903	2nd Lieut.	

Name.	Appointed to 9th Lancers.		Remarks.
	Date.	Rank.	
Marshall, Henry	April 1860	Captain	Exchanged from 17th Lancers with Captain Goldie.
Marshall, John	Feb. 1875	Riding-Master	Private, 18th Hussars, Feb. 1863.
Martin, William	April 1794	Cornet	
Martin-Bell, A. G.	April 1856	Lieut.	Transferred from Cape Mounted Rifles.
Martin-Bell, Edward H.	March 1878	2nd Lieut.	Died 1882.
Mathew, William	April 1799	Cornet	
Maunsell, E. H.	1886	Paymaster	
Maxwell, Richard	June 1797	Cornet	
Maxwell, Benjamin	May 1797	Assist.-Surgeon	
Mayne, ——	May 1794	Cornet	
M'Andrew, John	Feb. 1810	Assist.-Surgeon	
M'Calmont, Hugh	June 1865	Cornet	
M'Causland, Marcus M.	Nov. 1792	Chaplain	
M'Dermott, ——	August 1806	Lieut.	
M'Dowall, Andrew	Feb. 1765	Surgeon	
M'Dowall, George	April 1816	Cornet	
M'Duffie, Donald	Feb. 1813	,,	
M'Farlan, John	Dec. 1874	Paymaster	
M'Farlane, Francis	Jan. 1843	Cornet	Promoted 1st Dragoon Guards, 1855.
M'Farlane, Ronald	March 1880	2nd Lieut.	
M'Innis, E. B.	Oct. 1875	Lieut.	From 21st Hussars. Private, 16th Lancers, March 1865. Vice Lieut. Wheeler, retired.
M'Mahon, Thomas W.	April 1842	Captain	Appointed from 16th Lancers.
M'Nevin, D. Fitzgerald	May 1842	Cornet	
M'Queen, ——	June 1806	,,	
Meiklam, John	Dec. 1826	,,	
Meiklam, Robert	May 1835	,,	
Melin, George R.	Nov. 1830	Surgeon	
Melville-Whyte, J.	Dec. 1817	Cornet	
Menzies, John	Sept. 1826	,,	
Mills, Robert	Oct. 1850	,,	Private, 9th Lancers, April 1829.
Milly (ey), Nicholas	July 1715	Chaplain	
Minchin, John	June 1811	,,	
Mocher, Flower	April 1775	Colonel Comdg.	
Molesworth, The Viscount	May 1732	,,	
Montagu, The Hon. Oliver	July 1863	Cornet	
Montgomery, Hugh	1733	Captain	Died 1741.
Montgomery, Wm. Stone	Dec. 1771	Cornet	
Montgomery, Thomas	May 1822	Lieut.	
Montmorency, R. H. de	Jan. 1811	Maj. & Lt.-Col.	
Moore, Hugh	June 1739	Cornet	
Moore, R. St. Leger	April 1867	,,	Transferred to 12th Lancers, 1875.
Morgan, Anthony	May 1720	,,	
Morland, Charles	Feb. 1810	Major	
Morland, Henry C.	1889	,,	Exchanged from 5th Lancers with Major Scott-Chisholme.
Morres, Redmond	Jan. 1811	,,	
Morrogh, Alexander	June 1856	Cornet	
Moutray, James	Oct. 1746	,,	
Musgrave, J. Knightley	Nov. 1820	,,	
Need, Samuel	Jan. 1837	Colonel Comdg.	
Needham, William	Jan. 1798	Captain	
Neilson, George M.	August 1900	2nd Lieut.	
Nelthorpe, Henry	June 1846	Lieut.	Exchanged with Lieut. Studdert from the 75th Foot.
Newcomen, Brabazon	July 1794	Cornet	
Newport, Simon G.	June 1813	Lieut.-Colonel	
Nicholson, Gilbert T.	Feb. 1838	Cornet	Exchanged to 13th Light Dragoons with Capt. Deacon.
Nicholson, Eugene Henry	1888	Riding-Master	
Noel-King, The Hon. Lionel F.	August 1885	Lieut.	
Norris, Robert	May 1816	Cornet	
Norton, J.	Oct. 1814	Vet.-Surgeon	
Nuton, Thomas W.	Never	joined	

Name.	Appointed to 9th Lancers.		Remarks.
	Date.	Rank.	
O'Connor, James	Feb. 1815	Surgeon	
Odlum, George	March 1795	Cornet	
Orde, Charles	April 1799	Captain	
Ormerod, Oliver	Sept. 1858	Cornet	Promoted 75th Regiment, 1861.
Ormsby, Thomas	March 1760	,,	
Ormsby, William	June 1766	,,	
Ormsby, Arthur	April 1780	Captain	
Ormsby, Philip	May 1781	Cornet	
Ouvry, Henry A.	Oct. 1854	Major	Exchanged with Major Foster from the 99th Regiment.
Palairet, Charles H.	Dec. 1865	Cornet	
Paliser, Thomas Howe	Oct. 1776	Capt. & Lieut.	
Parker, Henry Thomas	June 1803	Cornet	
Parker, George	1895	Riding-Master	
Patrickson, Christopher	June 1794	Lieut.	Exchanged to 43rd Lt. Infantry.
Paulet, Charles W.	March 1860	Captain	Exchanged from 7th Hussars with Captain Shaw.
Payne, Sir Charles, Bart.	July 1811	Cornet	
Payne, Henry L.	August 1852	,,	Appointed to 5th Dragoon Guards, 1856.
Peach, George	June 1809	Surgeon	
Peacock, Wilkinson	June 1806	Cornet	
Pearson, H.	1715	Lieut.-Colonel	
Pelham, The Hon. Victor R.	Nov. 1886	Lieut.	
Pemberton, Roger	July 1715	Cornet	
Pennefather, Kingsmill	March 1800	Lieut.	
Percy, William	August 1867	Cornet	Private, 4th Light Dragoons, April 1842.
Perry, John P.	April 1803	Lieut. & Adjt.	
Peters, John Godfrey	Dec. 1808	Captain	
Pigot, George	April 1806	,,	
Pollok, John F.	Jan. 1899	2nd Lieut.	
Pomeroy, Arthur	Feb. 1775	Capt. & Lieut.	
Porter, H. E.	July 1817	Cornet	
Porter, Thomas	April 1833	,,	
Power, K. Manley	Feb. 1837	,,	Exchanged to 16th Lancers with Captain Yule, 1847.
Powis, Richard J.	April 1756	,,	
Powys, The Hon. Charles	April 1842	Lieut.	From 16th Lancers.
Pratt, Edward	July 1847	Captain	Exchanged from 16th Lancers with Captain Macartney.
Preston, Nathaniel	1726	Cornet	
Preston, George	1749	Chaplain	
Preston, Arthur	July 1766	Lieut.-Colonel	
Preston, Nathaniel	Jan. 1769	Cornet	
Preston, The Hon. Edward F.	Dec. 1864	,,	
Pretor, S. Ashton	Dec. 1856	,,	
Prince, William	Sept. 1733	Surgeon	
Pringle, Joseph	Sept. 1739	Cornet	
Purdon, George	Oct. 1812	Assist.-Surgeon	
Quaile Vere, Essex	Feb. 1788	Chaplain	
Ramsden, H. J.	Nov. 1826	Captain	Exchanged from 15th Hussars with Captain The Hon. G. Vaughan.
Ratcliffe, Thomas H.	Jan. 1852	Paymaster	Exchanged with Barlow.
Rathbone, ——	Nov. 1801	Cornet	
Read, E. Rudston	Oct. 1841	Captain	Exchanged from 13th Hussars with Captain A. Thomson.
Reade, George	Nov. 1749	Colonel Comdg.	
Reynard, Francis H.	Nov. 1882	Lieut.	Transferred from 15th Foot. Promoted to Troop 7th Hussars. Exchanged back to Regiment with Captain G. Evans.
Renant or Renaut, Peter	Nov. 1719	Cornet	
Reynolds, Francis	July 1719	,,	Was a Quartermaster, July 1715.
Reynolds, Francis	Sept. 1746	Captain	
Ricardo, William P.	March 1878	2nd Lieut.	Killed in action, 11 Dec. 1879.
Rich, Charles D.	June 1851	Cornet	

Name.	Appointed to 9th Lancers.		Remarks.
	Date.	Rank.	
Richards, W. F.	Jan. 1845	Cornet	To 17th Lancers, 1848.
Richardson, H. J.	Sept. 1811	Lieut.	
Riddell, Robert	Dec. 1775	Cornet	
Rind, John	Oct. 1822	Cornet and Riding-Master	Was a half-pay Ensign, "York Chasseurs," 1819. A private, 9th Lancers, 1793.
Robarts, James W.	May 1842	Lieut.	Exchanged from 10th Hussars with Lieut. Hyder.
Robertson, Robert	Feb. 1810	,,	
Robertson, James	Sept. 1829	Vet.-Surgeon	Exchanged from the Regiment, 1842, with Vt.-Surg. Johnston.
Robinson, Robert	May 1809	Cornet	
Rocksavage, George, Earl of	Oct. 1901	2nd Lieut.	
Rolfe, Joseph	Oct. 1806	Lieut.	
Rollo, Andrew	May 1740	Cornet	
Rose, J. R. Holden	April 1842	Captain	Promoted into 17th Lancers, Sept. 1857.
Rosslyn, The Earl of	August 1801	Colonel Comdg.	At date of appointment, was General Sir James St Clair Erskine, Bt.
Rumley, Randall	Dec. 1824	Cornet	
Russell, Christopher	1724	Lieut.	
Russell, James	Dec. 1781	Cornet	
Russell, John	April 1858	,,	
Russell, The Hon. George	Oct. 1871	Sub-Lieut.	
Russell, James W. H.	March 1891	2nd Lieut.	
Sampson, William J.	July 1803	Captain	
Sarel, Henry A.	July 1847	Cornet	Promoted 17th Lancers, 1857.
Saunders, Richard	April 1756	,,	
Saunders, Anderson	Nov. 1772	,,	
Saunders, Thomas A.	Feb. 1861	,,	
Saurin, James	August 1788	Chaplain	
Savage, Patrick	May 1801	Cornet	
Savile, William	Jan. 1863	,,	
Schroder, William H. von	Dec. 1889	2nd Lieut.	
Scott, Anthony	Jan. 1760	Cornet	Retired 1766.
Scott, John	April 1842	Lieut.-Colonel	On augmentation, from 4th Light Dragoons.
Scott, Henry	May 1848	Cornet	Promoted 18th Light Dragoons, 1858.
Seggie, Samuel	May 1861	Quartermaster	
Shaw, Thomas	Sept. 1809	Lieut.	
Shaw, Richard	May 1846	Cornet	Private, 17th Lancers, 1823. Transferred from 3rd Light Dragoons. Exchanged to 7th Hussars with Captain Paulet, 1860.
Shaw, A. Cooper	April 1857	Vet.-Surgeon	Exchanged with Hurford.
Shawe, R. Fleetwood-	June 1823	Cornet	
Shearburn, Egremont E.	May 1874	Lieut.	
Shiell, William J.	June 1826	Assist.-Surgeon	
Sidley, Richard	June 1800	Quartermaster	
Skeffington, William J.	August 1770	Lieut.	
Skene, Andrew P.	August 1791	Captain	
Sleigh, James Wallace	August 1839	Colonel Comdg.	
Smales, Thomas	Sept. 1858	Paymaster	Exchanged with Mahon.
Smith, Henry	July 1715	Captain	
Smith, Tennison	April 1771	Cornet	
Smith, George Gordon	Dec. 1814	,,	
Smith, Donald S.	Feb. 1856	Assist.-Surgeon	To 6th Dragoon Guards, 1857.
Smith-Kincaid, Thomas M.	Oct. 1894	2nd Lieut.	
Smoke, Thomas	May 1794	Cornet	
Smyth-Skeffington, Geoffrey H.	May 1895	2nd Lieut.	
Somerset, Charles Henry	Nov. 1821	Captain	
Southwell, The Hon. Henry	April 1742	,,	Vice Lord Charles Hay. An M.P., December 1745.
Spalding, John E.	August 1827	Cornet	Exchanged from 3rd Hussars with Lord Fincastle.
Spicer, J. W. Gooch	April 1836	,,	Exchanged to 3rd Dragoon Guards, 1841.

Name.	Appointed to 9th Lancers.		Remarks.
	Date.	Rank.	
Spottiswoode, Andrew	April 1834	Cornet	From 14th Foot. Exchanged to 1st Dragoon Guards with Major Allen, 1851.
Spry, George F.	June 1864	Assist.-Surgeon	
Stackpoole, George B.	June 1814	Cornet	
Stanley, Herbert F.	Feb. 1887	2nd Lieut.	Died of wounds, South Africa, 3rd May 1900.
Stannus, Thomas	Sept. 1811	Lieut.	
Staples, John	Oct. 1762	Captain	Resigned 1764.
Staples, William Deane	July 1794	Cornet	
Starkey, Thomas S.	Jan. 1858	,,	
Staunton, Walter G.	April 1842	Assist.-Surgeon	
Steele, A. Frederick	June 1850	Captain	By exchange from 98th Regiment.
Sterne, Enoch	April 1756	Cornet	Was Quartermaster in the Regiment, 1748, or earlier.
Stevenson, Norman S.	March 1861	Riding-Master	
Steward, Henry H.	April 1850	Cornet	Promoted 2nd West India Regiment, 1856.
Stewart, Archibald	August 1841	Assist.-Surgeon	
Stirling, Gilbert C.	June 1862	Cornet	To Royal Horse Guards.
Stirling, J. Gordon	July 1895	2nd Lieut.	Killed at polo, Sialkote, May 1902.
Strickland, Miles	April 1766	Cornet	
Stuart, Simon H.	Nov. 1814	Captain	
Studdart, Charles F.	April 1842	Cornet	Exchanged to 75th Foot with Lieut. Nelthorpe.
Sutherland, John	Nov. 1837	,,	Exchanged to 56th Foot, 1841.
Sutherland, Robert	May 1848	,,	Promoted 4th Light Dragoons, 1851.
Swetnam or Swetenham	1748	,,	A Quartermaster, 1744.
Swift, Richard	Sept. 1806	Assist.-Surgeon	
Talbot, Francis	April 1793	Cornet	
Talbot, The Hon. Richard	Sept. 1865	,,	Lord Talbot de Malahide.
Taylor, James	1725	,,	
Tennison, William	Feb. 1801	,,	
Theobald, Stanley R.	Sept. 1897	2nd Lieut.	
Thomas, Bartholomew B.	May 1808	Lieut.	
Thompson, ——	1728	Cornet	
Thomson, John Anstruther	Feb. 1836	,,	Exchanged to 13th Light Dragoons, 1841, with Captain Rudston Read.
Thonger, R. Freer	June 1857	,,	Private, 9th Lancers, Feb. 1841.
Thorpe, William S.	August 1896	2nd Lieut.	Killed by an accident, bicycling, Rosboom, Natal, Nov. 1897.
Thursby, Piers	Sept. 1856	Cornet	
Thynne, Lord John	Dec. 1886	Lieut.	Killed, by an accident, at York, May 1887.
Tibboe, Thomas	March 1754	Surgeon	Died 1764.
Tomkyns, W.	Nov. 1855	Cornet	Never joined.
Tooth, Douglas K.	August 1900	2nd Lieut.	
Tribe, Edward G.	Oct. 1871	Sub-Lieut.	
Trower, Edward S.	Sept. 1825	Cornet	
Trower, Frederick C.	May 1842	Lieut.	From 16th Lancers. Exchanged to 50th Foot with Capt. Fawcett, 1855. Judge-Advocate-General in India, being the first officer of the Royal Service appointed to this post.
Trower, Charles J.	Feb. 1875	,,	Transfer from 4th Hussars.
Trower, Arthur S.	Nov. 1899	2nd Lieut.	
Trueman, C. J.	April 1831	Cornet	Never joined.
Tuite, Marcus A.	April 1771	Captain	
Turner, Edmund J.	May 1842	Lieut.	Exchanged from 4th Dragoon Guards.
Tuson, Edward B.	Oct. 1851	Assist.-Surgeon	Promoted 12th Lancers, 1854.
Uniacke, Richard	Nov. 1790	Cornet	
Upton, Lewis	March 1827	Lieut.	Exchanged from 4th Dragoons with Lieut. and Adjt. Wright.
Upton, Roger D.	April 1855	,,	Exchanged from 1st West India Regiment with Lieut. King.

Name.	Appointed to 9th Lancers.		Remarks.
	Date.	Rank.	
Vandelure or Vandeleur, Richd.	July 1757	Cornet	
Vaughan, The Hon. George .	May 1821	,,	
Vesey, George . . .	June 1825	,,	
Vesey, The Hon. Eustace .	April 1873	Sub-Lieut.	Died Nov. 1886.
Vincent, Sir Francis, Bart. .	Dec. 1818	Cornet	
Wallace, William . . .	May 1791	Surgeon	
Wallace, James Maxwell .	August 1805	Cornet	
Waller, Robert . . .	Nov. 1756	,,	
Waller, Arthur . . .	Feb. 1759	,,	
Walpole, The Hon. George .	April 1780	,,	
Warde, George . . .	May 1778	Captain	
Waring, John	Sept. 1803	Cornet	
Warnes, Jacob or John . .	July 1715	Lieut.	
Warren, Francis . . .	August 1808	,,	
Webster, Edward . . .	May 1720	Cornet	
Webster, Henry . . .	June 1811	,,	
Wedderburn, J. K. . . .	April 1842	Lieut.	From 2nd Life Guards.
Weeks, Nathaniel . . .	Sept. 1825	Cornet	
West, David	Feb. 1796	,,	
Westenra, Henry . . .	Sept. 1741	Lieut.	
Westenra, Joseph . . .	March 1788	Major	
Westenra, The Hon. D. W. .	Dec. 1872	Sub-Lieut.	
Westenra, The Hon. R. H. .	Jan. 1880	2nd Lieut.	Private, 4th Dragoon Guards, Sept. 1875. Died Dec. 1880.
Westropp, Henry . . .	August 1798	Cornet	
Whalley, Thomas P. . .	May 1832	,,	
Whatman, H. A. . . .	Jan. 1832	,,	Died August 1836.
Wheeler, A. L. . . .	April 1868	,,	Retired October 1875.
White, Samuel. . . .	March 1801	,,	
Whitley, Henry . . .	April 1759	Colonel Comdg.	
Whitney, Edward . . .	July 1715	Lieut.	
Wicklow, The Earl of . .	Jan. 1865	,,	
Wildman, Thomas . . .	Oct. 1816	Major	
Wilford, Richard R. . .	August 1782	Captain	
Wilkins, John West . .	August 1790	Cornet	
Wilkinson, Alfred E. . .	Feb. 1848	,,	To 7th Hussars, on exchange with Captain Bulkeley, 1859.
Williams, William . . .	Jan. 1797	,,	
Williams, Percy B. . . .	Sept. 1822	,,	Died 1876.
Williams, A. C. . . .	June 1825	,,	
Williams, Norton T. . .	March 1842	Lieut.	
Williams, Edward . . .	June 1871	Captain	
Williamson, J. Clark . .	August 1838	Quartermaster	Private, 9th Lancers, Feb. 1810. Died April 1841.
Willis, Frederick . . .	Nov. 1825	Cornet	
Willis, John George . .	March 1850	,,	Died October 1858.
Wilson, Henry C. . . .	Jan. 1806	Lieut.	
Willoughby, Francis D. . .	Sept. 1836	Cornet	Died April 1846, at Meerut.
Willoughby, The Hon. Ernest	July 1867	,,	
Willoughby, The Hon. C. H.	Dec. 1884	Lieut.	
Witherington, William . .	July 1715	,,	
Witherington, Edward . .	July 1791	Captain	
Witherington, Henry . .	Jan. 1797	Cornet	
Wood, John	Feb. 1796	,,	
Wood, Arthur	July 1841	Surgeon	Exchanged with Ingham.
Wood, James A. . . .	Sept. 1877	Vet.-Surgeon	
Wood, Herbert F. . . .	May 1901	2nd Lieut.	
Wright, John	March 1789	Cornet	Resigned January 1790.
Wright, John	June 1811	Cornet and Adjt.	
Wright, Richard . . .	April 1818	Cornet	
Wright, Thomas . . .	August 1825	Quartermaster	Private, 9th Lancers, Oct. 1805.
Wynne, Owen	July 1715	Colonel	This officer formed the Regiment.
Wynne, Owen	July 1715	Cornet	
Wynne, John	June 1761	Lieut.-Colonel	
Wynn, Robert V. . . .	Oct. 1899	2nd Lieut.	
Young, A. Semple . . .	Nov. 1849	Cornet	
Yule, Robert A. . . .	July 1847	Captain	Exchanged from 16th Lancers with Captain Power.
Zarzhetsky, Karl C. . .	Jan. 1902	2nd Lieut.	

APPENDIX IX.

APPEN

THE REGIME

ANNUAL LIST OF OFFI

The rank given is that which officers held the first time they attended a regimental dinner. The letter given by past officers of the regiment to those serving and about to embark

Rank.	Name.	1866.	20 May 1867.	1868.	24 May 1869.	1870.	1871.	1872.	1873.	1 June 1874.	1875.*	1875.	29 May 1876.	28 May 1877.	3 June 1878.	26 May 1879.
Captain	Abadie, H. R.	..	..	No record.	..	..	..	..	..	..	G	..	..	..	..	..
Mr.	Adams, J. P.	..	..		..	..	..	..	..	..	G	..	..	..	..	..
,,	Agnew, C.	..	..		P	..	..	..	..	..	..	..	..	..	..	..
Captain	Albert Victor, H.R.H. Prince	..	..		..	..	..	..	..	..	..	..	..	..	..	..
Mr.	Allhusen, F. H.	..	..		..	..	..	..	..	..	..	..	..	..	..	..
,,	Alston, H. R.	..	..		..	..	..	..	..	..	..	..	..	..	..	..
Major	Apperley, W. H.	..	..		..	..	..	..	..	..	..	..	..	..	..	..
Captain	Arbuthnot, W.	..	..		..	..	..	..	..	..	..	..	..	..	P	..
Captain	Barlow, Sir Morrison, Bart.	..	..		..	..	..	..	..	..	..	..	..	..	..	..
Mr.	Barrow, L.	..	..		P	..	..	..	..	..	..	..	..	..	..	..
,,	Barton, R.	..	..		..	..	P	P	P	P	H	P	P	P	..	..
Captain	Bashford, C. B.	..	..		P	..	P	P	P	P	H	P	P	P	P	P
Mr.	Beale-Browne, D. J.	..	..		..	..	..	..	..	..	..	..	..	..	..	..
,,	Bell, E. F.	..	..		..	..	..	..	..	..	..	..	..	..	..	..
Lieutenant	Bentinck, Lord Charles	..	..		..	..	..	..	..	..	..	..	..	..	..	..
Cornet	Beresford, Lord Wm. (V.C.)	..	..		..	..	P	P	P	P	G	..	..	..	..	..
Mr.	Bishop, C.	..	..		..	..	..	..	..	..	..	..	..	..	..	..
Doctor	Breslin, W. J.	P	P		P	..	..	..	..	..	..	..	..	..	..	..
Mr.	Brooke, V. R.	..	..		..	..	..	..	..	..	..	..	..	..	..	..
,,	Broughton, R.	..	..		..	..	..	..	..	..	..	..	..	..	..	..
Captain	Bushman, H. A.	..	..		..	..	..	P	P	P	G	..	..	P	P	P
Mr.	Butson, S. G.	..	..		..	..	..	P	..	..	G	..	..	..	..	..
,,	Cameron, C. St. C.	..	..		..	..	..	..	..	..	..	..	..	..	..	..
,,	Campbell, D. G.	..	..		..	..	..	..	..	..	..	..	..	..	..	..
,,	Candy, H.	P	P		P	..	P	..	..	..	H	..	P	P	P	P
Lieutenant	Carleton, The Hon. D. P.	..	..		..	..	..	..	..	..	..	..	..	..	..	..
Captain	Cave-Browne-Cave	..	..		..	..	..	P	P	..	H	P	..	..	P	P
Mr.	Cheeseman, E.	..	..		..	..	..	P	..	..	..	..	..	..	..	..
,,	Chisholme, J. Scott-	..	..		..	..	..	P	..	..	..	P	..	..	..	..
Captain	Clark, P.	P	..		P	..	..	P	..	..	H	P	..	P	P	..
,,	Clayton, W. C.	..	P		..	..	P	P	..	P	G	..	..	..	..	..
,,	Cleland, R. S.	..	..		..	..	..	P	..	P	G	P	P	..	..	..
Doctor	Clifford, J.	..	..		..	..	..	..	P	..	H	..	..	..	..	..
Mr.	Colvin, F. F.	..	..		..	..	..	..	..	..	..	..	..	..	..	..
Lieutenant	Compton, Lord Douglas	..	..		..	..	..	..	..	..	..	..	..	..	..	..
Lieut.-Gen.	Davie, Sir F.	P	..		..	..	..	..	..	..	..	..	..	..	..	..
Mr.	De Laski, J. L.	..	..		..	..	..	..	..	..	..	..	..	..	..	..
	Digby, Lord	P	P		P	..	P	P	P	..	..	..	..	..	..	..
Colonel	Drysdale, W.	P	P		P	..	P	P	P	..	..	..	P	P	P	P

DIX IX.

NTAL DINNER.

CERS DINING, 1866-1903.

"P" against an officer's name indicates that he attended the dinner of that year. * This was a dinner for India. The letter "H" indicates the Hosts; "G," the Guests.

1880.	1881.	22 May 1882.	21 May 1883.	26 May 1884.	1 June 1885.	24 May 1886.	1887.	28 May 1888.	3 June 1889.	2 June 1890.	25 May 1891.	30 May 1892.	29 May 1893.	4 June 1894.	27 May 1895.	1 June 1896.	31 May 1897.	23 May 1898.	29 May 1899.	1900.	1901.	2 June 1902.	25 May 1903.
..	P	P	..	..	..	P	..	P	P	P	P	P	P	P	P	P	P	P	P	South African War. No dinner held.	South African War. No dinner held.	P	P
..	P	..	..	..	..	..	..	..	..	..	..	..	..	..	..	..	..	..	..			..	..
..	..	..	..	..	..	..	..	..	..	..	..	..	..	..	..	..	..	..	..			..	..
..	..	..	..	..	..	..	..	P	P	P	..	..	..	..	..	..	..	..	..			..	..
..	..	..	..	..	..	..	..	..	..	..	..	..	..	..	..	..	..	..	..			..	..
..	..	..	..	..	..	..	..	..	..	..	..	..	P	P	P	P	P	P	..			P	..
..	..	..	..	..	..	..	..	P	..	..	P	..	..	..	..	..	..	..	..			..	P
..	P	..	..	..	..	..	..	P	..	P	..	..	..	..	..	..	..	..	..			..	..
..	P	..	..	..	P	P	..	..	..	..	..	..	..	..	..	..	..	..	..			..	..
..	..	P	..	..	..	..	..	..	..	..	..	..	..	..	..	..	..	..	..			..	..
..	..	..	..	..	..	..	..	..	..	..	..	..	..	..	..	..	..	..	..			..	..
..	..	..	..	..	..	..	..	..	..	..	..	..	..	..	..	..	..	..	..			..	..
P	P	P	P	..	P	P	..	P	P	P	..	P	P	P	P	P	P	P	P			P	..
..	..	..	..	..	..	..	..	..	..	..	..	P	P	..	P	P	..	..	P			..	..
..	..	..	..	..	..	..	..	..	..	..	..	..	..	..	P	P	..	..	P			..	P
..	..	..	..	..	..	..	..	..	..	P	P	..	..	..	P	P	..	..	P			P	P
..	..	..	..	..	P	..	..	P	..	P	..	P	..	P	..	..	P	P	P			..	..
..	..	..	..	P	..	P	..	P	P	P	P	..	P	..	P	P	P	P	P			P	P
..	..	..	..	..	..	..	..	..	..	..	..	..	..	..	..	..	..	..	..			..	..
..	..	..	..	..	..	..	..	..	..	..	..	..	..	..	..	P	..	..	..			..	..
..	..	..	P	..	..	..	..	..	..	..	..	..	..	..	..	..	..	..	..			..	..
..	..	P	..	..	..	..	..	P	..	P	P	P	P	P	..	P	P	P	P			P	P
..	..	..	..	..	..	..	..	..	..	..	..	..	..	..	..	..	..	..	..			..	P
..	..	..	..	..	..	P	..	..	P	P	P	..	..	..	..	..	..	..	..			..	..
..	..	..	..	..	..	..	..	..	P	..	P	P	P	P	P	P	..	..	P			P	..
P	..	P	..	P	..	P	..	..	..	P	..	..	P	..	..	P	P	P	..			P	..
..	..	..	..	..	..	..	..	..	..	..	..	..	..	..	..	..	..	..	..			P	..
P	P	P	..	P	..	..	..	..	..	P	..	P	..	..	..	..	P	P	..			..	..
..	..	..	..	..	..	..	..	..	..	..	..	..	..	..	..	..	..	..	..			..	..
P	P	P	..	..	..	P	..	P	..	..	..	..	..	P	..	..	..	..	..			..	..
..	..	..	..	..	..	..	..	..	..	..	..	..	..	..	..	..	P	P	P			..	..
..	..	..	..	..	..	..	..	..	..	..	..	..	..	..	..	..	..	..	..			..	..
..	..	..	..	..	..	..	..	..	..	..	..	..	..	..	..	..	..	..	..			..	..
..	..	..	..	..	..	..	..	..	..	..	..	..	..	..	..	..	..	..	..			..	..
..	..	..	..	P	..	P	..	..	P	P	P	P	P	P	..	P	..	..	P			..	..
..	..	..	..	..	..	P	..	P	P	P	..	..	..	..	..	P	..	..	..			P	..
..	..	..	..	..	..	..	..	..	..	..	..	..	..	..	..	..	..	..	..			..	..
..	..	..	..	..	..	..	..	P	P	P	..	..	..	..	..	..	..	..	..			..	..
..	..	..	..	..	..	..	..	..	..	..	..	..	..	..	..	..	..	..	..			..	..
P	P	P	P	P	..	P	..	P	P	P	P	P	P	P	P	P	P	P	P			..	..

THE REGIMENTAL DINNER—ANNUAL LIST

Rank.	Name.	1866.	20 May 1867.	1868.	24 May 1869.	1870.	1871.	1872.	1873.	1 June 1874.	1875.*	1875.	29 May 1876.	28 May 1877.	3 June 1878.	26 May 1879.
Mr.	Duckett, J. S.	..	..		..	..	..	..	..	..	..	..	..	..	..	..
,,	Duff, F. W.	..	..		..	..	..	..	..	..	..	..	..	..	..	..
Major.-Gen.	Elliott, Sir A.	..	..		..	..	..	..	..	..	..	..	..	..	..	..
Captain	Ellis, F.	P	P		P	..	P	P	P	P	H	P	P	..	..	..
Mr.	Ellison, G. P.	..	..		..	..	..	..	..	..	..	..	..	..	..	..
Captain	Erskine, W. M.	..	..		P	..	P	..	P	..	H	P	P	P	P	P
Mr.	Evans, G. A. P.	..	..		..	..	..	..	..	..	G	..	..	..	..	..
Captain	Evans, J.	..	P		P	..	P	P	..	P	H	..	P	P	P	P
Major	Fiennes, The Hon. I. de V.	..	P		..	..	P	P	P	P	H	..	..	..	..	..
Mr.	Fiennes, H. T. W.	..	..		..	..	..	..	..	..	..	..	..	..	..	..
,,	Fife, W. H.	..	..		..	..	P	P	..	..	G	..	..	P	P	P
,,	Floyd, C.	..	..		..	..	..	..	..	..	..	..	..	..	..	P
Major	Follett, S. W.	..	..		..	..	..	..	..	..	..	..	..	..	..	..
Mr.	Forbes, J.	..	..		..	..	..	..	..	..	..	..	..	..	..	..
,,	Forsythe, J.	..	..		..	..	..	..	..	..	..	..	..	..	..	..
Major	Francis, T. J.	..	..		..	..	..	..	..	..	..	..	P	..	..	..
Mr.	Fuller, G.	..	..		..	..	..	..	..	..	G	..	..	..	P	..
Captain	Gaskell, H. B.	..	..		..	..	..	P	..	P	H	..	P	..	P	P
Mr.	Gerard, F.	..	..		P	..	P	P	P	..	..	..	P	P	..	P
Major	Goldie, J.	..	..		..	..	..	..	..	..	..	..	..	P	..	..
Captain	Goldsworthy, R. T.	..	..		..	..	..	..	..	..	H	..	..	..	..	..
Mr.	Gough, B.	..	..		..	..	..	..	P	P	G	..	..	..	..	..
,,	Green, P.	..	..		P	..	P	P	..	P	G	P	P	P	P	P
,,	Grissell, F.	P	P		P	..	P	P	..	..	G	..	..	P	P	P
Lieut.-Gen.	Grant, Sir J. Hope	P	P		P	..	P	P	P	P	..	..	..	..	..	..
Lieut.-Col.	Grant, F. R.	..	..		..	..	..	..	..	..	..	P	..	..	..	..
Doctor	Grant, M.	..	..		..	..	..	..	..	..	H	..	..	..	..	..
Captain	Hankey, W.	..	..		..	..	..	P	..	..	..	..	..	..	..	..
Major	Hardy, John	..	..		..	..	..	P	..	..	..	..	P	P	..	..
Mr.	Henry, F.	P	P		P	..	P	P	P	P	H	..	P	P	P	P
,,	Henry, Gilbert	..	..		..	.	..	..	..	..	..	..	..	..	..	..
,,	Herbert, F.	..	..	No record.	P	..	P	P	..	P	H	P	P	P	..	P
,,	Hopkins, H. C.	..	..		..	..	..	..	..	..	..	..	..	..	..	..
Captain	House, P.	..	..		P	..	..	P	P	..	..	..	..	..	..	..
Mr.	Hunter, J.	..	..		..	..	..	..	..	..	..	..	..	..	P	..
Captain	Isacke, F. J.	..	..		..	..	..	..	..	..	..	..	..	..	P	..
Mr.	Jenner, W. K.	..	..		..	..	..	..	..	..	..	..	..	..	..	..
Major	Johnson, C.	P	P		P	..	P	..	..	..	..	..	..	..	..	..
Mr.	Johnson, E. P.	..	..		..	..	..	..	..	..	..	..	..	..	..	..
Major	Jones, A. S. (V.C.)	..	..		..	..	P	..	..	P	..	..	..	P	..	..
Mr.	Kincaid-Smith, T.	..	..		..	..	..	..	..	..	..	..	..	..	..	..
Lieutenant	King-Noel, The Hon. L.	..	..		..	..	..	..	..	..	..	..	..	..	..	..
Mr.	Lamont, J. H.	..	..		..	..	..	..	..	..	..	..	..	..	..	P
Lieutenant	Lascelles, The Hon. C.	..	..		..	..	..	..	..	..	G	..	..	..	P	P
Mr.	Lawrell, W.	P	..		..	..	P	..	..	..	..	..	..	..	..	P
Captain	Legge, The Hon. H.	..	..		..	..	..	..	..	P	G	..	..	P	P	P
Colonel	Little, Archibald	P	P		..	..	P	P	P	P	H	P	P	P	P	P
Mr.	Little, M. O.	..	..		..	..	..	..	..	..	..	..	..	..	..	..
,,	Longman, G.	..	..		..	..	..	..	..	..	H	..	..	..	..	..
Captain	Lund, F. T.	..	..		..	..	..	..	..	..	..	..	..	..	..	..
Mr.	Mackenzie, J. A. Stewart-	..	..		..	..	..	P	..	P	G	..	..	P	..	..
,,	Mackenzie, E. P.	..	P		P	..	P	P	P	P	H	..	P	P	P	P
,,	Mackintosh, R.	..	P		..	..	..	..	..	..	..	..	..	..	..	..
Colonel	Madocks, J. E.	P	..		P	..	P	P	P	..	..	..	P	P	P	..
Major	Mahon, M. H.	..	..		..	..	..	..	P	P	H	..	P	P	P	P
Doctor	Malcolm, J.	..	..		..	..	P	..	..	..	G	..	..	..	..	..
Mr.	Martin-Bell, A. G.	..	..		..	..	..	..	..	..	..	..	..	..	..	..
Captain	Marshall, H.	..	P		..	..	P	P	P	P	G	..	..	..	P	..
,,	M'Calmont, H.	..	..		..	..	P	..	P	P	H	..	..	..	..	..
Mr.	M'Farlane, R.	..	..		..	..	..	..	..	..	..	..	..	..	..	..
,,	M'Innis, E. B.	..	..		..	..	..	..	..	..	..	..	..	..	..	..
Major.-Gen.	M'Mahon, Sir T.	..	..		..	..	..	..	..	..	H	..	..	..	..	..
Mr.	Moore, R.	..	..		..	..	P	P	..	..	G	..	P	P	P	..
Major	Morland, H.	..	..		..	..	..	..	..	..	..	..	..	..	..	..
Captain	Morragh, A.	..	P		..	..	..	..	..	..	..	..	..	..	..	..
Mr.	Palairet, C.	P	..		..	..	P	P	P	P	G	P	P	P	P	P
Lieutenant	Pelham, The Hon. V.	..	..		..	..	..	..	..	..	..	..	..	..	..	..
Mr.	Percy, W.	..	..		..	..	..	..	..	..	G	..	..	..	..	..
Major.-Gen.	Porter, Thomas	P	P		..	..	..	..	..	..	..	..	..	..	..	..
Captain	Power, R.	..	P		P	..	..	P	..	..	..	..	..	..	..	..

OF OFFICERS DINING, 1866-1903—*continued.*

1880.	1881.	22 May 1882.	21 May 1883.	26 May 1884.	1 June 1885.	24 May 1886.	1887.	28 May 1888.	3 June 1889.	2 June 1890.	25 May 1891.	30 May 1892.	29 May 1893.	4 June 1894.	27 May 1895.	1 June 1896.	31 May 1897.	23 May 1898.	29 May 1899.	1900.	1901.	2 June 1902.	25 May 1903.
..	..	..	..	..	..	..	..	..	..	..	..	..	..	..	..	..	..	..	..	South African War. No dinner held.	South African War. No dinner held.	P	..
..	P	..	..	..	..	P	..	P	..	..	..	..	..	..	..	..	P	..	..			P	P
..	..	..	..	..	..	..	..	..	..	..	..	..	..	..	..	..	..	..	..			P	..
..	P	P	P	P	P	P	..	P	P	P	..	..	..	..	..	..	..	..	..			..	..
..	..	..	..	..	..	..	..	..	P	P	..	P	P	..	P	P	P	P	..			..	..
P	P	P	P	P	..	P	..	P	P	P	..	P	..	P	P	P	P	P	P			..	..
P	..	..	..	..	..	P	..	..	..	P	..	P	..	..	..	..	P	P	P			P	P
P	..	P	P	P	P	P	..	P	P	..	P	P	P	P	P	P	..	..	P			..	..
..	..	..	..	..	..	..	..	..	..	..	..	..	..	..	..	..	..	..	..			..	..
..	..	..	..	..	..	..	..	P	P	P	..	..	P	..	P	..	..	..	..			..	..
P	P	P	P	P	P	P	..	P	P	P	P	P	P	..	P	P	P	P	..			P	P
P	P	..	P	P	P	..	..	..	..	..	..	..	..	..	..	..	..	..	..			..	..
..	..	..	..	..	..	..	..	..	..	..	..	..	..	..	P	P	..	..	..			..	..
..	..	..	..	..	..	..	..	..	..	..	..	..	..	P	P	..	P	P	P			P	P
..	..	..	..	..	..	P	..	..	..	..	..	..	..	..	..	..	..	..	..			..	..
..	..	..	..	..	..	P	..	..	..	..	..	..	..	..	..	..	..	..	..			..	..
..	..	..	..	..	..	..	..	P	P	P	P	P	P	P	..	..	..	..	..			..	..
..	P	P	P	P	..	P	..	P	..	P	P	P	P	P	P	P	P	P	P			P	P
..	P	P	..	..	..	..	..	..	..	..	..	..	..	..	..	..	..	..	..			..	..
..	..	..	..	..	..	..	..	..	..	..	..	..	..	..	..	..	..	..	..			..	..
..	..	P	..	..	..	..	..	..	..	..	..	..	P	..	..	..	..	..	..			..	..
..	P	..	..	P	P	P	..	..	P	..	..	..	..	..	P	P	..	..	P			P	P
P	..	P	..	P	..	..	..	P	P	P	P	P	P	P	P	P	P	P	P			P	..
P	P	P	..	..	..	..	..	..	..	..	..	..	..	..	..	..	..	..	..			..	P
..	..	..	..	..	..	..	..	..	..	..	..	..	..	..	..	..	..	..	..			..	..
..	..	..	..	..	..	..	..	..	..	..	..	..	..	..	..	..	..	..	..			..	..
P	..	..	..	..	..	..	..	..	..	..	..	..	..	..	..	..	..	..	..			..	..
..	..	..	..	..	..	..	..	..	..	..	..	..	..	..	..	..	..	..	..			..	..
..	..	..	..	..	..	P	..	..	..	..	..	..	..	..	..	..	..	..	..			..	..
P	P	P	P	..	P	P	..	P	P	..	P	P	P	P	P	P	P	P	P			P	P
..	..	..	..	..	..	..	..	P	P	P	P	..	..	..	P	P	..	..	P			P	P
..	..	P	..	..	..	P	..	..	P	P	P	..	P	..	..	P	P	P	P			..	P
..	P	..	P	P	..	..	..	..	..	..	..	..	..	..	..	..	..	..	..			..	..
..	..	..	..	..	..	..	..	..	..	..	..	..	..	..	..	..	..	..	..			..	..
..	P	..	P	P	..	P	..	..	..	..	P	..	..	..	..	..	..	..	..			..	..
..	..	..	..	..	..	..	..	..	..	..	..	..	..	..	..	..	..	..	..			..	..
..	..	..	..	..	..	P	..	P	P	P	P	P	P	P	P	P	P	P	..			..	P
..	..	..	..	..	..	..	..	..	..	..	..	..	..	..	..	..	..	..	..			..	..
..	..	P	..	..	P	P	..	..	..	..	..	..	..	..	..	..	..	..	..			..	..
..	..	..	P	P	..	P	..	P	P	P	P	P	P	P	P	..	..	..	..			..	..
..	..	..	..	..	..	..	..	..	..	..	..	..	..	..	P	P	..	..	..			..	..
..	..	..	..	..	..	P	..	P	P	P	..	..	..	..	P	P	P	P	P			..	P
..	P	P	P	..	..	P	..	P	P	P	..	..	P	P	P	P	P	P	P			..	..
P	P	P	P	P	..	..	..	..	..	..	..	..	..	..	..	..	..	..	..			..	..
..	..	..	..	..	..	..	..	..	..	..	..	..	..	..	..	..	..	..	..			..	..
..	P	P	..	..	..	P	..	..	P	P	P	P	..	P	P	P	P	P	P			P	P
P	..	P	P	P	P	P	..	P	P	P	..	..	..	..	..	..	..	..	..			..	..
..	..	P	P	..	..	P	..	P	P	P	P	P	..	P	P	..	..	..	..			..	..
..	..	..	..	..	..	..	..	..	..	..	..	..	..	..	..	..	..	..	..			..	..
..	..	..	..	..	..	..	..	..	..	..	..	P	P	..	P	P	..	..	..			P	..
..	..	..	P	..	..	P	..	..	..	..	P	P	P	P	P	P	P	P	..			..	P
P	P	P	P	P	P	P	..	P	P	P	..	P	P	P	P	P	P	P	P			P	P
..	..	..	..	..	..	..	..	..	..	..	..	..	..	..	..	..	..	..	..			..	..
P	..	P	..	..	..	P	..	..	P	..	..	..	..	..	..	..	..	..	..			..	..
P	P	P	P	P	..	P	..	P	P	P	P	..	..	..	..	..	..	..	..			..	..
..	..	..	..	..	..	..	..	..	..	..	..	..	..	..	..	..	..	..	..			..	..
..	..	P	..	..	..	..	..	..	..	..	..	..	..	..	..	..	..	..	..			..	..
P	P	P	P	P	P	P	..	P	P	P	..	P	..	P	P	P	P	P	P			P	P
..	..	..	P	P	..	P	..	..	..	P	..	..	P	..	..	P	..	..	..			..	..
..	..	..	..	P	..	P	..	P	P	..	..	..	..	..	..	..	..	..	..			..	P
..	P	..	..	..	..	P	..	P	..	P	..	P	..	..	P	P	P	P	..			P	P
..	..	..	..	..	..	P	..	..	..	..	..	..	..	..	..	..	..	..	..			..	..
P	..	..	P	P	P	P	..	P	P	P	P	P	..	P	P	P	P	P	P			P	..
..	..	..	..	..	..	..	..	..	..	..	..	P	P	P	..	..	..	..	P			P	..
P	..	P	..	..	..	P	..	P	..	P	..	..	..	..	..	..	..	..	..			..	..
P	P	..	P	..	..	..	..	P	P	..	P	P	P	P	P	P	P	P	P			..	..
..	..	..	..	..	..	..	..	P	P	P	P	P	P	..	P	P	P	P	P			P	P
..	..	..	..	..	..	..	..	..	..	..	..	..	..	..	..	..	..	..	..			..	..
..	..	..	..	..	..	..	..	..	..	..	..	..	..	..	..	..	..	..	..			..	..
..	..	P	..	P	P	P	..	P	..	..	..	..	..	..	..	..	..	..	..			..	..

THE REGIMENTAL DINNER—ANNUAL LIST

Rank.	Name.	1866.	20 May 1867.	1868.	24 May 1869.	1870.	1871.	1872.	1873.	1 June 1874.	1875.*	1875.	29 May 1876.	28 May 1877.	3 June 1878.	26 May 1879.
Colonel	Powys, The Hon. C.	..	P		P	..	P	P	P	P	H	P	P	..	P	P
Captain	Preston, The Hon. E.	..	..		P	..	P	P	P	P	H	..	P	P	P	..
,,	Pretor, S. A.	P	P		P	..	P	P	P	P	H	..	..	P	P	P
Mr.	Read, Rudston, E.	..	..		..	..	..	..	..	..	H	..	..	..	..	..
,,	Reynard, F. H.	..	..		..	..	..	..	..	..	..	..	..	..	..	..
,,	Ricardo, W. P.	..	..		..	..	..	..	..	..	..	..	..	..	P	..
Major	Rich, C. D.	..	..		P	..	P	P	P	P	H	P	P	P	P	P
Captain	Richards, W. F.	..	..		..	..	..	..	..	..	..	..	..	..	..	..
Lieut.-Col.	Rose, H.	P	..		P	..	P	P	P	P	..	..	..	..	P	P
	Rossmore, Lord	..	..		..	..	..	..	..	..	H	..	..	P	..	..
Lieutenant	Russell, The Hon. G.	..	..		..	..	..	P	P	..	..	..	..	..	..	..
Doctor	Ruttledge, J.	P	..		P	..	P	P	..	..	..	..	..	..	..	..
Lieut.-Col.	Sarel, H.	P	P		P	..	..	P	..	P	H	..	P	P	P	P
Captain	Savile, W.	..	..		..	..	P	..	P	P	H	P	..	P	P	P
Mr.	Schröder, W. H. von	..	..		..	..	..	..	..	..	..	..	..	..	..	..
Major	Scott, H.	..	..		P	..	P	..	P	P	..	..	P	P	..	P
,,	Shawe, R. Fleetwood-	P	..		..	..	..	P	..	..	..	..	..	..	..	..
Mr.	Shearburn, E.	..	..		..	..	..	..	..	..	G	..	..	P	..	..
Major	Spicer, J. W.	P	P		P	..	P	P	P	P	..	..	P	P	P	..
Doctor	Spry, G. F.	..	..		P	..	P	P	P	P	H	..	P	P	P	P
Mr.	Smyth-Skeffington, T.	..	..		..	..	..	..	..	..	..	..	..	..	..	..
,,	Stanley, H. F.	..	..		..	..	..	..	..	..	..	..	..	..	..	..
Captain	Starkey, T. S.	P	P	No record.	P	..	P	P	P	P	..	P	P	P	P	P
Colonel	Steele, A. F.	..	P		P	..	..	P	..	P	..	..	P	P	..	P
Mr.	Stirling, G.	P	P		P	..	..	P	P	P	H	..	P	P	P	P
,,	Stirling, J. Gordon-	..	..		..	..	..	..	..	..	..	..	..	..	..	..
Cornet	Talbot, The Hon. R.	..	..		..	..	..	..	..	..	..	..	..	P	..	..
Mr.	Thomson, J. Anstruther	..	..		..	..	..	..	..	..	..	..	P	..	P	P
Captain	Thonger, R.	..	..		P	..	..	P	P	..	..	..	..	..	P	..
,,	Trower, F. C.	..	..		P	..	P	P	..	..	H	..	P	..	..	..
Mr.	Trower, C. J.	..	..		..	..	..	..	..	..	..	..	..	..	..	..
Captain	Thursby, P.	P	P		P	..	P	P	P	P	H	P	P	P	P	P
,,	Vesey, G.	..	..		P	..	P	P	..	..	..	..	..	..	..	..
Lieutenant	Vesey, The Hon. E.	..	..		..	..	..	..	P	P	G	..	..	P	P	P
	Wedderburn, J. K.	..	..		..	..	..	..	..	..	..	..	..	..	..	..
Captain	Weeks, N.	..	P		..	..	P	P	P	P	..	P	..	..	..	..
	Westenra, The Hon. D. W.	..	..		..	..	..	..	P	..	..	..	..	..	..	..
Mr.	Wheeler, A.	..	..		P	..	P	P	P	..	G	P	P	P	P	P
Captain	Wilkinson, A.	..	..		..	..	..	..	..	..	..	..	..	P	P	P
,,	Williams, P. B.	..	..		..	..	..	P	..	..	..	..	..	..	..	..
,,	Williams, E.	..	..		..	..	..	P	..	..	..	..	..	..	..	..
Cornet	Willoughby, The Hon. E.	..	..		..	..	P	P	..	..	G	P	..	P	P	P
Lieutenant	Willoughby, The Hon. C.	..	..		..	..	..	..	..	..	..	..	..	..	..	..
Mr.	Wood, J. A.	..	..		..	..	..	..	..	..	..	..	..	..	..	..

OF OFFICERS DINING, 1866-1903—*continued.*

1880.	1881.	22 May 1882.	21 May 1883.	26 May 1884.	1 June 1885.	24 May 1886.	1887.	28 May 1888.	3 June 1889.	2 June 1890.	25 May 1891.	30 May 1892.	29 May 1893.	4 June 1894.	27 May 1895.	1 June 1896.	31 May 1897.	23 May 1898.	29 May 1899.	1900.	1901.	2 June 1902.	25 May 1903.
P	P	P	P	P	P	P	..	P	P	..	..	P	P	P	..	..	..	..	..	South African War. No dinner held.	South African War. No dinner held.	..	..
P	P	P	P	P	P	P	..	P	P	P	P	P	P	P	..	P	P	P	..			..	..
P	..	..	..	..	..	..	..	..	..	..	..	..	..	..	..	..	..	..	..			..	..
..	..	..	..	..	..	..	..	..	..	..	..	..	..	..	..	..	..	..	..			..	..
..	..	..	..	..	..	P	..	P	..	P	P	P	P	P	P	P	P	P	P			P	P
..	..	..	..	..	..	..	..	..	..	..	..	..	..	..	..	..	..	..	..			..	..
P	P	P	P	P	P	P	..	P	P	P	P	P	P	..	..	..	..	..	..			..	..
P	P	..	..	..	..	..	..	..	..	..	..	..	..	..	..	..	..	..	..			..	..
..	..	..	..	..	..	..	..	..	..	..	..	..	..	..	..	..	..	..	..			..	..
P	..	..	..	..	..	..	..	..	..	..	..	P	..	..	..	P	..	..	..			..	..
..	..	..	..	..	..	..	..	..	..	..	..	..	..	..	P	..	P	P	..			..	..
..	..	..	..	..	..	..	..	..	..	..	..	..	..	..	..	..	..	..	..			..	..
P	P	P	P	P	..	P	..	..	..	..	..	..	..	..	..	..	..	..	..			..	..
P	P	P	P	P	P	P	..	P	P	P	P	P	P	P	P	P	P	P	P			P	..
..	..	..	..	..	..	..	..	..	..	P	..	P	P	P	P	P	P	P	P			P	..
P	P	P	P	..	P	P	..	..	..	..	..	..	..	..	..	..	..	..	..			..	..
..	..	..	..	..	..	..	..	..	..	..	..	..	..	..	..	..	..	..	..			..	..
..	..	..	..	..	..	..	..	..	..	..	..	..	..	..	..	..	..	..	..			..	..
P	..	..	..	..	..	..	..	..	..	..	..	..	..	..	..	..	..	..	..			..	..
P	P	P	P	P	P	P	..	P	P	P	P	P	P	P	P	P	P	P	P			..	..
..	..	..	..	..	..	..	..	..	..	..	..	..	..	..	..	P	..	..	..			P	P
..	..	..	..	..	..	..	..	P	P	..	P	P	P	P	P	P	P	P	..			..	..
P	P	P	P	P	P	P	..	P	P	P	P	P	..	..	..	..	..	..	..			..	..
P	P	P	P	..	..	..	..	..	..	..	..	..	..	..	..	..	..	..	..			..	..
P	P	P	..	P	P	P	..	P	..	..	..	..	..	P	..	P	P	P	P			P	P
..	..	..	..	..	..	..	..	..	..	..	..	..	..	..	..	P	..	..	P			..	..
..	..	..	..	..	P	..	..	..	..	..	..	P	..	..	..	..	..	..	..			..	..
..	P	P	..	..	..	..	..	..	..	..	..	..	..	..	..	..	..	..	..			..	..
P	..	P	P	P	..	P	..	P	..	..	..	..	..	..	..	..	..	..	..			..	..
..	..	P(?)	..	..	..	..	..	..	..	..	..	..	..	..	..	..	..	..	..			..	..
..	..	P(?)	P	..	..	..	..	P	..	P	P	P	P	..	P	P	P	P	P			P	P
P	P	P	P	P	P	P	..	P	P	P	P	P	P	P	P	P	P	P	..			..	..
..	..	..	..	..	..	..	..	..	..	..	..	..	..	..	..	..	..	..	..			..	..
P	..	..	..	P	P	P	..	..	..	..	..	..	..	..	..	..	..	..	..			..	..
..	..	..	..	..	..	P	..	..	..	..	..	..	..	..	..	..	..	..	..			..	..
..	..	..	..	..	..	..	..	..	..	..	..	..	..	..	..	..	..	..	..			..	..
..	..	..	..	..	..	..	..	..	..	..	..	..	..	..	..	..	..	..	..			..	..
P	P	P	P	P	P	P	..	P	P	..	P	P	P	P	P	P	P	P	P			P	P
P	P	P	P	P	P	P	..	P	P	P	P	P	..	..	..	..	..	..	..			..	..
..	..	..	..	..	..	..	..	..	..	..	..	..	..	..	..	..	..	..	..			..	..
..	..	..	..	..	..	..	..	..	..	..	..	..	..	..	..	..	..	..	..			..	..
..	P	P	P	P	..	..	..	P	..	..	..	..	P	..	..	..	..	..	..			P	..
..	..	..	..	..	..	P	..	P	..	P	P	..	..	..	P	..	..	..	..			..	..
..	..	..	..	..	..	P	..	P	..	..	P	..	..	..	..	..	..	..	..			..	..

INDEX.

PAGE

Abadie, Lieut. E. H., A.D.C. to General French . . 113
Accident, horse and man fall down a well . . . 60
horse, officer, and man fall down a well . . . 65
Adams, the Rev. J., awarded the Victoria Cross . . . 88
Adjutants, roll of, 1742-1904 . 183
Afghan War, start for . . 82
"Afghanistan" (1878-80), word to be borne on appointments 93
Agra, battle of 60
Albert Victor, H.R.H. Prince, 10th Hussars, attached to regiment 98
Allhusen, Lieut. F., A.D.C. to General Tucker . . . 113
Ambrose, Sergeant, awkward position of 111
Ankers, Sergeant-Major, plucky behaviour of . . . 124
Army Lists, Annual, 1715-1903 App. VII.
Aroyo de Molino, French troops surprised there . . . 25
Artillery, Royal, officers and men serve in 55
complimentary orders, &c., therewith 56

Ballymore Eustace, engagement at 16
Bamian Road, action near . . 85
Band, two cream-coloured colts for use of 35
one cream-coloured filly for use of 96
Bareilly, capture of . . . 74
Battles, roll of 149
Baxter, Quartermaster W., death of 144
Bell, Captain E. F., staff officer to Colonel Hunter-Weston . 133

PAGE

Bell-Martin, picquet under, saved the guns 54
gun and limber incident . . 69
Belmont, battle of . . . 109
Beresford, Lord William, awarded the Victoria Cross . . 83
dinner to, at Calcutta, when leaving India . . . 101
Bloemfontein, surrender of; the regiment selected for Lord Roberts's escort. . . 116
Boolundshuhur, action at . . 59
Bouwers Hork, capture of commando near . . . 134
Brandfort, engagement near, gallantry of men . . . 118
Brevet promotion, Afghan War . 93
South African War . . 139
Brigade, the 3rd Cavalry, composition of, November 1900. 127
officers of, dine at Bloemfontein 127
Bronkhorst Spruit, attempt to blow up bridge over . . 120
Brooke, Lieut. Victor, narrow escape of 108
Brown, Colonel John, biography of 168
Budlee-ki-Serai, engagement at . 50
Buenos Ayres, expedition to . 20
Bushman, Major-General Henry A., biography of . . 179
Butson, Captain, death of . . 88

Call, the regimental . . . 152
Campbell, Captain David, narrow escape of 109
Campbell's Cavalry Brigade, February 1858 . . . 71
Candahar-Kabul, march to . . 90
Candahar, battle of . . . 92
Cap, field-service, new pattern of 101

Carbines, application for, by Major A. Little, 1858 . 46
issue of 82
new pattern of Martini-Metford 100
new pattern of Martini-Metford 101
new pattern of magazine . 102
Carlow, attempted taking of the town 16
Castle Comer, engagement near . 17
Casualties, Indian Mutiny, officers and men 75
South African War, officers and men 141
Cavagnari, embassy to Kabul under, massacred . . 83
Cavalry, "Malleson" on their work round Lucknow . 64
brigades, February 1858 . 71
division, Hope Grant in command 71
division, February 1900, composition of . . . 112
Cawnpore, battle of . . . 64
Charasia, action of . . . 84
action of, 25 April 1880 . 89
word to be borne on appointments 93
Chillianwallah, action of, etc. . 42
word to be borne on appointments 46
Chisholme, Colonel J. J. Scott-, death of 107
Ciudad Rodrigo, retirement on . 27
Civil power, in aid of, Wales . 98
Clayton, Captain W. C., death of 82
Cleland, Colonel, death of, field force order on . . . 91
Clothing, warrant concerning, 1747 10
Collar chains, abolished . . 100
Colonels of the regiment, roll of, 1715-1904 164
Cope, Brigadier-General Sir John, biography of . . . 168
Columns, mobile, all arms . 127
to act with Colonel White . 129
to act with Colonel Maxwell . 129
to act with Colonel Scobell . 132
Colvin, Major F., appointed staff officer 129
command of column . . 130
Commanding officers, roll of, 1715-1904 180
Cost, of a dragoon regiment, 1717 7
Cricket, 9th L. *v.* R.A., 1857 . 67
Crofts, Colonel James, biography of 166
Cruppers, use of, discontinued . 97
Cypher monogram, origin and date of 37

De Laski, Lieut. J., death of . 99
Delhi Field Force, formation of . 47
Field Force, first move of . 49
final assault on . . . 57-58
ex-King, escorted to Mirzapore 75
and "Lucknow," words to be borne on appointments . 79
Demobilization, order for, 1902 . 136
Desertion, deserter reward, 1715 4
Despatches, names mentioned, South African War . . 138
Diamond Hill, action of . . 122
Dilkoosha, engagement near . 72
Dinner, annual, established 1860 79
regimental, officers dining at App. IX.
Distances marched during South African War . . . 137
Distinguished Service Order, recipients of . . . 138
Conduct Medal, recipients of . 138
Division, Cavalry, Hope Grant in command, 1858 . . 71
Dorn Kop, rearguard action near 129
Dragoons, regiments of, raised 1715, career of . . 3
styled 9th Dragoons . . 8
constituted Light Dragoons . 14
Dress, alterations in, 1715-1903 App. V.
Drysdale, Lieut.-General Sir William, biography of . 178
Duckett, Lieut., captures Field-Cornet Prætorius . . 116
Durand, Lieut., badly wounded . 121
Durbar, at Rawal Pindi, 1885 . 94
at Delhi, 1902 . . . 144

Edenburg, escort of wrecklings to 130
Ellison, Captain, death of . . 121
Engagements and operations, South Africa . . . 137
Enslin, engagement at . . 109
Erskine, Sir James St. Clair (*see* Rosslyn) 172
Escorts, Royal, 1889 . . 99
Escort, to Lord Roberts, entry of Bloemfontein . . . 116
Establishments, of regiment, 1715-1903 . . . 153
Evans (Bashi), a narrow escape . 50
alligator catching . . . 55
and a bee party . . . 63

Exeter, regiment entertained by inhabitants thereof . . 77

Farewell Order, by the Viceroy, India, 1859 . . . 76
Farewell Order, by the Commander-in-Chief, India, 1859 75
messages, from Lord Kitchener and General French . . 143
Field-Cornet, capture of, by Lieutenants Sadleir-Jackson and Trower 123
Forsythe, Quartermaster John, retirement of . . . 101

Gauntlets, worn on mounted parades only . . . 78
Gogra, passage of river . . 75
Goojerat, battle of . . . 45
word to be borne on appointments 46
Goorsaigunj Bridge, action at . 69
Gordon-Stirling, Lieut., wounded 110
at Ospoort Nek . . . 131
accident to, death of . . 144
Grangue, De, Colonel Henry, biography of 168
Grant, Hope, Sir James, nearly shot on court-martial parade 42
narrow escape, Subzee-Mundee Road 52
his native orderlies . . 54
joins the Agra column, and smart work connected therewith 61
Corporal Caine incident . . 69
meeting with Dr. Russell . 72
biography of 175
Grant, Lieut. Frank, accident to his charger . . . 63
gharri accident to . . . 68
Grave, of Lieut.-Colonel Fullerton, at Sirinigar . . . 46
Graves, of Lieut.-Colonel R. Yule and Captain Clayton . . 52
Guidons, history of . . . 149
description of the regiment's . 150
use of, discontinued . . 151
Gundamuk, treaty of . . . 83
Gungaree (or Khasganj), action of 68
Gymkana, at Graff Reinet . . 132

Hartigan, Sergeant, gallant conduct of, etc. . . 61, 74
Hawkes and Masky, appointed cap-makers . . . 38
Head-ropes, cotton, taken into use 100
Hockey, matches at . . . 111
Honywood, Major-General Philip, biography of . . . 170
Hore, Colonel, relief of his column 124
Horse furniture, alterations in App. V.
Horses, to buy in England, 1720 . 8, 10
cream-coloured, for the band 35, 96
state of, February 1900 . . 114
to hand to 5th Lancers, 1902 . 136
Hout Kraal, engagement near . 136

India, arrival in, 1875 . . 81
departure from, 1885 . . 95
return to, 1898 . . . 103
departure from, 1899 . . 104
return to, 1902 . . . 144
Indian Mutiny, Delhi Field Force, formation of . . 47
Ireland, regiment proceeds to, 1717 7
quaint billeting order . . 7
Irish Rebellion, 1798 . . 15
last of 9th officers, who served in it 18

Johnston, Major-General Sir James, biography of . . 171
Jones, Lieut. A. S., wounded twenty-two places . . 60
Jorden, Colonel John, biography of 169
Jubilee, service and review, York, 1887 97
Jummoo, escort at, for the Prince of Wales 81

Kabul, embassy to, massacred . 83
capture of the city . . . 85
Candahar, march from . . 90
"Kabul" (1879) and "Kandahar" (1880), words to be borne on appointments . . . 93
Kallee Nuddee, action near, quaint incident . . . 62
Kila Karez, action at . . . 86
Kila Kazi, action of, losses in 86, 87
Kilcullen, engagement at . . 15
Kimberley, relief column, composition of . . . 108
relief of, by cavalry division . 113
relief of, charge by 9th and 16th Lancers . . . 114
Klip Drift to, General French on conduct of 3rd Cavalry Brigade 133

Kitchener's Hill, engagement near 115
Kumbda Kothi, regiment's final engagement, Mutiny . . 75
Kurrum Valley Field Force, 1st squadron with, composition of 83

"Lancers, Exercise of," first taught, 1816 . . . 28
Lancers, 9th, 12th, 16th, and 23rd Light Dragoons, so called 29
brigade of, reviewed, 1819 . 29
designated "Queen's Royal". 34
9th, description of, during Mutiny (Malleson) . . 63
9th, description of, during Mutiny (Captain O. Jones, R.N.) 66
Lance, successful use of . . 115
Lance-flag, for use, entry of Pretoria 121
Lances, use of, discontinued . 127
use of, abolished . . . 145
Leighlin, small engagement at . 17
Leeufontein, Boshof Road, engagement on . . . 114
Light Dragoons, formed so . 14
Limburgher, capture of, by Compton and Bell . . 133
Lists of officers by seniority, 1715-1903 . . . App. VII.
Little's Cavalry Brigade, 1858 . 71
Little, Colonel A., badly wounded 72
Colonel M. O., appointed a Brigadier 123
Colonel M. O., wounded . 125
General Sir Archibald, biography of 177
Lotter's commando, capture of . 134
Loughborough, Lord, biography of 182
Lucknow, relief of . . . 64
Luiperskop, action near . . 108
Lund, Captain, commands yeomanry column . . . 132

Macdonald, The Hon. A., sad, but gallant end of . . 130
Magersfontein, battle of . 110-11
Marches, daily, Kabul to Candahar 91
regimental, names of . . 152
Meanggunge, fort of, capture . 71
Medals and clasps, South African War. 138
for distinguished service, Afghan War . . . 93
Medals and clasps, Afghan War, Kabul-Candahar march . 94
Mercer, Sergeant, gallantry of . 119
Merida, engagement near . . 26
Merves, Van de, laager captured 113
Mess, sergeants', established . 36
sergeants', abolished . . 36
officers', monthly order for stores, 1857 . . . 53
Messages, farewell, on leaving South Africa . . . 143
Mocher, General Flower, biography of 171
Modder River Cup, won by Colonel Little's "Orcano". 111
Molesworth, The Viscount, biography of 166
Monogram, cypher, origin and date of 37
Moosa Bagh, action at . . 73
Morland, Colonel Charles, retirement of 30
Colonel Charles, biography of 181
Muster-roll certificate, 1743 . 9
Mutiny, Indian, casualties during 75
Myburg's Farm, engaged near . 132

Need, Lieut.-General Samuel, biography of . . . 174
New Ross, attack on the town of 16

Officers, first, in the regiment . 4
distribution of, outbreak Indian Mutiny 48
wounded, South African War. 141
left for duty, South Africa, regiment to India . . . 143
landed in India, 1902 . . 144
annual seniority lists of, 1715-1903, . . . App. VII.
alphabetical list of. . App. VIII.
Order, farewell, by Commander-in-Chief, India, 1859. . 75
farewell, by the Viceroy, India, 1859 76
to cavalry division, starting for Kimberley . . . 113
to cavalry division, after entry of Bloemfontein. . . 117
to the army, on entry of Pretoria 122
Organisation, squadron, established 100
Ospoort, engagement near. . 131

Paardeberg, surrender of Cronje. 116

Paardeplaats Farm, sharp fighting near 123
Patrickson, last 9th officer, served Irish Rebellion . . . 18
Pattiali, action near . . . 68
Pay, rates of, and allowances, 1757 11
rates of, and allowances, 1778 12, 13
letter concerning . . . 13
for six months, 1798 . . 18
Peninsula, campaign of, 1811 . 24
word to be borne on appointments 28
Plastron, front of tunic, origin of 163
Pollok, Lieutenant, death of . 121
Polo tournament, September 1900 125
Preston, battle of, 1715 . . 5, 6
Pretoria, entry of, by Lord Roberts 121
Promotion, brevet, for Afghan War 93
brevet, for South African War 139
special, non-com. officer and men 140
Punniar, action of, medal for, word to be borne on appointments . . . 40

Quaggafontein, engagement near 125
Quartermasters, roll of, 1800-1904 184
Quarters and service, list of, 1715-1904 . . 154-57

Rawal Pindi, Durbar at . . 94
Reade, Lieut.-General George, biography 169
Refreshment-room, regimental, establishment of . . 81
Regiment of Dragoons, cost of, 1717 7
Regimental staff, October 1899 . 107
call and marches . . . 152
Regiment, full colonels of, 1715-1904 164
commanding officers of, 1715-1904 180
Regimental dinner, officers dining at . . . . . App. IX.
Remounts, Argentine, very moderate 118
Revolvers, new pattern, 1887 . 98
new pattern, 1892 . . 100
Ribera, engagement near . . 27
Ridge, The, Delhi, capture of . 51
Rietfontein, outpost engagement near 126
Rifle, regiment armed with . 127
issue of magazine, Lee-Metford 145
Roberts, Lord, recommended by Hope Grant for the V.C. . 72
Rohilcund, operations in . 73, 74
Roodekop, engagement near . 118
Rosslyn, second Earl of, biography of 172
Royal Artillery, officers and men serve in 55
complimentary orders and letters therewith . . 56
Sabraon (Sobraon), battle of, medal for, word borne on appointments 41
Sadleir-Jackson, Lieut., badly wounded 121
Salute, awarded one, leaving India, 1859 . . . 76
Scheepers, escape of . . . 133
Seraighat Ferry, action at . . 66
Sergt.-Major Dunwoody, takes French colonel prisoner . 26
Service and quarters, 1715-1904 . . . 154-57
Shamsabad, action of . . 70
Sharkill, small engagement near . 17
Shatar-gardan, engagement at . 84
Sherpur, other squadrons join the 1st at 85
Siah Sung Heights, action near . 88
Signal Hill, ford near, incident at 107
Skeffington-Smyth, gallantry of . 125
Sleigh, General Sir James Wallace, biography of . . 174
Slik Spruit, engagement near . 128
South Africa, embarkation for, 1896 102
South African War, outbreak of . 104
regimental staff, commencement of 107
under fire, first time . . 108
engagements and operations . 137
South Africa, embarkation from . 143
officers left there, regiment proceeded to India . . 143
Springhaan Nek, De Wet escapes 129
Squadrons, officers serving with, June 1857 . . . 51
organisation of, established . 100
three service, and a reserve troop 103
re-formed into four . . 103
officers, &c. of, October 1899 . 107
regiment so weak, formed into two 122
regiment re-formed into three . 126

Squadrons appointed to mobile columns 127
Standards, history of 149
description of the regiment's . 150
use of, discontinued . . . 151
Standing Orders, first issued to the regiment 28
revised edition, 1819 . . 29
revised edition, 1832 . . 36
revised edition, 1860 . . 79
revised edition, 1886 . . 96
revised edition, 1902 . . 145
Stanley, Captain, death of . . 119
Stratford, near Carlow, engagement there 16
Subzee - Mundee Road, engagement at, death of Lieut.-Colonel Yule 52
Sword, new pattern, 1886 . . 96
new pattern, 1890 . . . 100

Tablet, memorial, St. James's Church, Graff Reinet . . 142
Thaba'nchu, outpost engagement near 119
Theobald, Lieut., prisoner at Pretoria, rejoins . . 121, 127
badly wounded 128
death of 133
Thorpe, Lieut. W. S., accident to, death of 103
Thynne, Lord John, accident to, death of 97
Transport ship, wreck of . . 23
Wardha, encounters severe gale 106
Troop, Colonel, Lieut. - Colonel, and Major, cease to command 19
Trumpeter Martin, French colonel surrenders to . . 25

Umballa, temporary return to . 74
return to, after Afghan War . 93

Veterinary Surgeon Landers, killed at Buenos Ayres . 23
Victoria Cross, Lieut. A. S. Jones, awarded 50
Captain Drysdale, recommended for 59
officers and men awarded, for Indian Mutiny 78
Lord William Beresford, awarded 83
Vinegar Hill, action at . . 17
Visch River Station, engagement near 133

Walcheren, expedition to, fever at 24
Welgelegen, 3rd Brigade joins main army at 119
Whitley, Lieut.-General Henry, biography of 171
Wreck, transport ship, *James and Rebecca* 23
Wynne's Dragoons, formation of, warrant for . . . 2
where raised 4
Wynne, Major - General Owen, biography of 165

THE END.

www.ingramcontent.com/pod-product-compliance
Ingram Content Group UK Ltd.
Pitfield, Milton Keynes, MK11 3LW, UK
UKHW021342130926
15127UKWH00011B/87

9 781843 422365